TECHNIQUES AND INSTRUMENTATION IN ANALYTICAL CHEMISTRY — VOLUME 27

UV-VISIBLE SPECTROPHOTOMETRY OF WATER AND WASTEWATER

TECHNIQUES AND INSTRUMENTATION IN ANALYTICAL CHEMISTRY

UV-VISIBLE SPECTROPHOTOMETRY OF WATER AND WASTEWATER

Edited by

O. Thomas

Observatoire de l'Environnement et du Développement Durable
Université de Sherbrooke
Sherbrooke, Québec
Canada

and

C. Burgess

Burgess Consultancy
Barnard Castle
Co. Durham, UK

ELSEVIER

AMSTERDAM • BOSTON • HEIDELBERG • LONDON • NEW YORK • OXFORD
PARIS • SAN DIEGO • SAN FRANCISCO • SINGAPORE • SYDNEY • TOKYO

Elsevier
Radarweg 29, PO Box 211, 1000 AE Amsterdam, The Netherlands
The Boulevard, Langford Lane, Kidlington, Oxford OX5 1GB, UK

First edition 2007

Library of Congress Cataloging-in-Publication Data
A catalog record for this book is available from the Library of Congress

British Library Cataloguing in Publication Data
A catalogue record for this book is available from the British Library

ISBN-13:   978-0-444-53092-9
ISBN-10:   0-444-53092-4
ISSN: 0167-9244

For information on all Elsevier publications
visit our website at books.elsevier.com

Printed and bound in The Netherlands

07 08 09 10 11   10 9 8 7 6 5 4 3 2 1

Working together to grow
libraries in developing countries

www.elsevier.com | www.bookaid.org | www.sabre.org

ELSEVIER   BOOK AID International   Sabre Foundation

PREFACE

The world of water and wastewater quality monitoring is beginning to gain significance. The development of more powerful analytical techniques and procedures in laboratories opens up the possibility of detecting a lot of emerging contaminants, while new screening tools are being developed for on-site measurement. One of the main reasons for this evolution is the European Water Framework Directive, obliging all member states to plan a monitoring programme. In this context, the use of simple, fast and relevant techniques such as UV spectrophotometry is to be considered.

UV-visible spectrophotometry is a mature analytical technique, from which numerous procedures have been developed, but is still poorly exploited. In fact, absorptiometry is the main tool, and a few methods are based on the study of the absorbance spectrum of water and wastewater. However, UV absorptiometry is still used as a detection technique for liquid chromatography and a coloured reaction is always needed for visible colorimetric methods.

Meanwhile, several important developments have been recently proposed either for instruments and accessories (for example, diode array spectrophotometers or optical fibre probes) or for spectra exploitation (for example, chemometric software). These improvements have led to a lot of applications, based on the exploitation of a large part of the UV spectrum.

Therefore, the aim of the book is to present practical UV spectrophotometry procedures, showing its interest for the direct examination of water and wastewater, and which can be handled with any PC-controlled spectrophotometer. From UV colorimetry to new multiwavelength approaches, water quality measurement can be carried out with a gain in time and simplicity.

Starting from basic considerations and general tools for spectral exploitation, different methods will be presented for the determination or estimation of the concentrations of organic and inorganic constituents. Moreover, some complementary measurements and tests will also be proposed.

A large part of the work is dedicated to applications dealing with natural water, urban and industrial wastewater, leachates and process control. These applications, original for most of them, result from more than 25 years of experience in the field of water and wastewater quality survey.

The last chapter presents various spectra of organic and inorganic constituents in water. This part of the book constitutes a very useful library of UV spectra in aqueous solution, giving the opportunity to the user to acquire spectral data for their own applications.

I would like to thank some people without whom this work would never have come into existence: Chris Burgess, my patient co-editor and contributor; Daniel Constant, who has transformed some of my scientific dreams into instruments; Derek Coleman, the quiet witness of a long process; Mireille Dussault, my assistant and corrector; all my co-workers contributing to this book; and, finally, Marie-Florence, Melissa, Camila, Anggi and David, to whom I dedicate this work.

Olivier Thomas
Sherbrooke, August 16, 2006

CONTENTS

6. Physical and Aggregate Properties 145
M.-F. Pouet, N. Azema, E. Touraud, O. Thomas

7. Natural Water 163
M.-F. Pouet, F. Theraulaz, V. Mesnage, O. Thomas

8. Urban Wastewater 189
O. Thomas, F. Theraulaz, S. Vaillant, M.-F. Pouet

S. Spinelli, C. Gonzalez, O. Thomas

UV-Visible Spectrophotometry of Water and Wastewater
O. Thomas and C. Burgess (Eds.)
© 2007 Elsevier. All rights reserved.

CHAPTER 1

The Basics of Spectrophotometric Measurement

C. Burgess

*Burgess Consultancy, 'Rose Rae', the Lendings, Startforth,
Barnard Castle, Co Durham, DL12 9AB, England*

1. INTRODUCTION

This book is concerned with the application of UV-visible spectrophotometry to the identification and determination of materials in a variety of water samples. The spectra included in this book form a resource that enables users to apply the technique to their own samples. However, in order to apply UV-visible spectrophotometry properly and reliably, we need to have some understanding of the principles and practices upon which it is based. The purpose of this short section is to cover the essential elements. The reader who wishes to explore the subject further may turn to the more detailed and specialised works referenced in the bibliography.

2. INTERACTION OF LIGHT AND MATTER

2.1. The electromagnetic spectrum

Spectroscopic processes rely on the fact that electromagnetic radiation (EMR) interacts with atoms and molecules in discrete ways to produce characteristic absorption or emission profiles. This is examined in more detail in Section 2.2. Before we can look into the origin of spectra, we have to look at some of the properties of EMR.

Our own ability to perceive colour is due to the human eye acting as a detector for EMR. The property of EMR that determines the range of colour perceived is wavelength. The part of the electromagnetic spectrum that the eye can detect is known, unsurprisingly, as the visible region. EMR may be simply represented as a sine wave. Wavelength, λ, is the distance between adjacent peaks or troughs. This is illustrated in Fig. 1.

Our ability to perceive colour depends on many factors. However, the interaction mechanism of EMR with matter is of major importance. These optical processes will be discussed in more detail in the next section. From a visual detection point of view, our ability to perceive different colours is dependent on the optical process involved, for example, if the light is absorbed or reflected by the observed object.

The wavelength, λ, of EMR can be expressed as a function of its frequency, ν, and the speed of light, c, by the following simple equation:

$$\nu = \frac{c}{\lambda} \qquad (1.1)$$

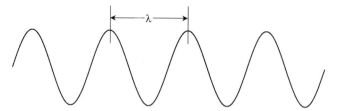

FIGURE 1. *Sine wave representation of electromagnetic radiation.*

Our eyes are not uniform with response to EMR in the visible region of the spectrum. They are most sensitive in the region of 600 nm. Figure 2 shows the relative sensitivity of the eye to visible light. This figure illustrates the importance of detector sensitivity and wavelength range for spectroscopic detectors. These and other instrument-related matters will be looked at in more detail in Section 2.

However, EMR behaves as a particle and as a wave (the dual nature of light); and the wavelength of such a particle, a photon, is related to energy by the equation

$$E = \frac{hc}{\lambda} 10^9 \tag{1.2}$$

where h is the Planck's constant (6.63×10^{-34} Js), c is the speed of light in vacuum (2.998×10^8 ms^{-1}), E is the energy of the photon and is the wavelength in nm.

The visible region of the electromagnetic spectrum constitutes but a tiny part, as can be seen from Fig. 3.

It is evident that there is an enormous span of energies, of over 18 orders of magnitude. The equation linking energy to wavelength is of fundamental importance in spectroscopy and will be discussed further in the next section.

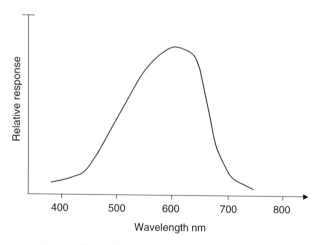

FIGURE 2. *The spectral sensitivity of the eye as a detector.*

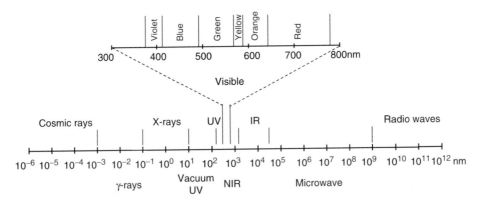

<small>FIGURE</small> 3. *The electromagnetic spectrum.*

2.2. The origin of spectra, absorption of radiation by atoms, ions and molecules

When a photon interacts with an electron cloud of matter, it does so in a specific and discrete manner. This is in contrast to the physical attenuation of energy by a filter that is continuous. These discrete absorption processes are quantised and the energies associated with them relate to the type of transition involved.

2.2.1. Fundamental processes

We can illustrate the process by way of a simple calculation using Equation (1.2). Assume that a photon of energy 8.254×10^{-19} joules interacts with the electron cloud of a particular molecule and causes promotion of an electron from the ground to an excited state. This is illustrated in Fig. 4. The difference in the molecular energy levels, $E_2 - E_1$, in the molecule corresponds exactly to the photon energy.

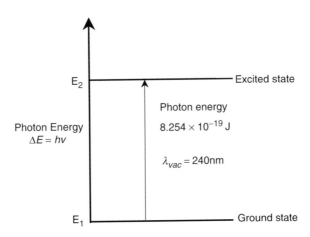

<small>FIGURE</small> 4. *Photon capture by a molecule.*

Converting this energy into wavelength reveals that this excitation process occurred at a wavelength of 240 nm. This is an electronic transition and is in the ultraviolet part of the spectrum. If this were the only transition that the molecule was capable of undergoing, it would yield a sharp single spectral line.

Molecular spectra are not solely derived from single electronic transitions between the ground and excited states. Quantised transitions do occur between vibrational states within each electronic state and between rotational sublevels. As we have seen, the wavelength of each absorption is dependent on the difference between the energy levels. Some transitions require less energy and consequently appear at longer wavelengths.

Considering a simplified model of a diatomic molecule, we might expect that our spectra would be derived from the three transitions between the ground and first excited state illustrated in Fig. 5. In practice, of course, even the simplest diatomic molecules have many energy levels resulting in complex spectra.

These processes and their consequences in observed spectra are discussed later in the book.

2.2.2. Optical processes in spectrophotometry

Thus far, we have only considered the absorption of energy by electronic and molecular transitions. When we make spectral measurements, it is necessary to consider other optical processes. This is particularly important for solution spectrophotometry.

When light impinges on a cuvette containing our molecule of interest (solute) in a solution (solvent), other optical processes do or can occur:

- Transmission
- Reflection

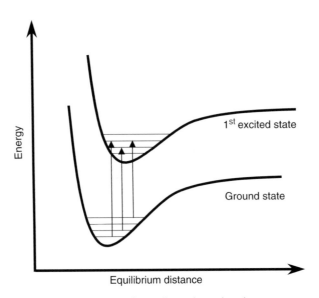

FIGURE 5. *Idealised energy transitions for a diatomic molecule.*

- Refraction
- Scattering
- Luminescence
- Chiro-optical phenomena

All these processes, together with instrumental effects, combine to distort or degrade the quality of the spectrum. The competent spectroscopist recognises these dangers and seeks to minimise their impact. More will be said about good spectroscopic practice in Section 3.1.

2.2.3. Chromophores

As we have seen from the previous sections, spectra are derived from quantised transition between energy states in atoms and molecules. The wavelengths at which these transitions occur are dependent upon the processes undergone. Hence, electronic transitions occur at higher energies (ultraviolet) than vibrational (infrared) or rotational ones (microwave). The molecular spectra observed in the UV-visible-NIR are a combination of these transitions. The intensity of the absorption is linked to the type of transition and the probability of its occurrence. Generally speaking, those transitions that are favoured in quantum mechanical terms exhibit more intense absorption bands.

Even simple molecules exhibit complex spectra in the UV portion of the spectrum. For example, benzene, as shown in Fig. 6, shows part of the vapour phase spectrum at a

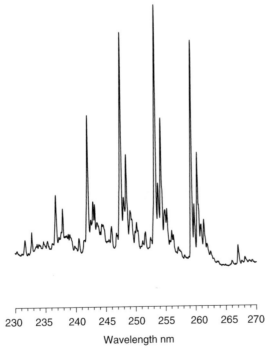

230   235   240   245   250   255   260   265 270

Wavelength nm

FIGURE 6. *Vapour phase spectrum of benzene in the region 230–270 nm.*

spectral bandwidth of 0.1 nm between 230 and 270 nm, associated with an electronic transition at about 260 nm. The amount of fine detail observed is the result of different transitions between vibration modes overlaid with even finer rotational structure.

The part of the molecule involved in these absorption processes is known as the chromophore. The spectra arising from different chromophores are the 'fingerprints' that allow us to identify and quantify specific molecules.

These chromophores are the basic building blocks of spectra and are associated with molecular structure and the types of transition between molecular orbitals. There are three types of ground state molecular orbitals—sigma (σ) bonding, pi (π) bonding and nonbonding (n)—and two types of excited state—sigma star (σ^*) and antibonding, pi star (π^*) antibonding—from which transitions are observed in the UV region. These are illustrated in Fig. 7.

These four transitions yield different values for ΔE and, hence, wavelength. Simple unconjugated chromophores can be characterised using these descriptors. Some examples are given in Table 1.

In addition, in the visible region, ligand field and charge transfer spectra are also observed. For further information regarding the origin of UV-visible spectra and the effect of conjugation on chromophores, Rao [3], Dodd [4] and Jaffé and Orchin [5] should be consulted.

Solution spectra, which are the concern of this book, are much less complex, particularly those involving polar solvents such as water or alcohols.

Figure 8 shows the solution spectrum of benzene in the nonpolar solvent hexane. Only the main vibrational fine structure is observed now due to solute–solvent interactions.

This type of 'fingerprint' is indicative of many benzenoid compounds, and the band intensities and positions are influenced by substituents. This topic will be discussed in more detail in Section 3 and illustrated in the spectral library. Hence, it is essential to control many parameters when measuring solution spectra including:

- pH
- Solvent purity and polarity
- Solute concentration
- Temperature
- Ionic strength

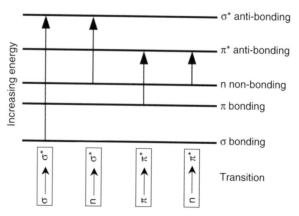

FIGURE 7. *Transitions between molecular orbitals.*

TABLE 1. *Examples of absorption maxima for isolated chromophores [1,2]*

Chromophore	Transition	Approximate wavelength of maximum absorption (nm)
$-\overset{\mid}{\underset{\mid}{C}}-\overset{\mid}{\underset{\mid}{C}}-$	$\sigma \rightarrow \sigma^*$	150
$-O-$	$n \rightarrow \sigma^*$	185
$-N\big\langle$	$n \rightarrow \sigma^*$	195
$-S-$	$n \rightarrow \sigma^*$	195
$\big\rangle C=O$	$\pi \rightarrow \pi^*$	170
	$n \rightarrow \pi^*$	300
$\big\rangle C=C\big\langle$	$\pi \rightarrow \pi^*$	170

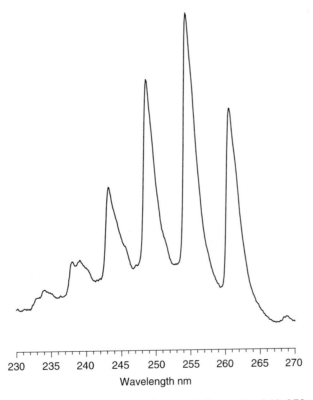

FIGURE 8. *Solution spectrum of benzene in hexane in the region 240–270 nm.*

2.3. Quantitative laws of the attenuation of light

Let I_0 be the intensity of a parallel beam of radiation of wavelength λ incident on and passing through a cuvette containing a solution of thickness b. Ignoring any losses from scattering or reflection, the emerging beam has been attenuated by the absorption process to an intensity I. This is illustrated in Fig. 9.

Note that for work of the highest accuracy, a single cuvette should be used for both the sample and reference solutions to ensure that the scattering and reflection losses are compensated for and any effects minimised.

The change in intensity of the incident beam dI caused by the thickness db of the absorbing solution is given by Lambert's law:

$$dI = -k_\lambda\, db \qquad (1.3)$$

where k_λ is a wavelength-dependent constant.

Rearranging and integrating between the limits of intensity from I_0 to I and path length from 0 to b, the equation becomes

In a similar manner, the change in intensity of the incident beam, dI, caused by the concentration increment of an absorbing material, dM, in the solution thickness, db, is given by Beer's law, where M is the concentration of the absorber in moles per dm^3:

$$dI = -k_\lambda\, dM \qquad (1.4)$$

where k_λ is another wavelength-dependent constant.

These two laws may be combined to give the familiar Beer–Lambert law:

$$\log_{10}\frac{I_0}{I} = \frac{k'_\lambda bM}{2.303} \qquad (1.5)$$

where the constant term $k'_\lambda/2.303$ is called the molar absorptivity, and left-hand side of the equation is the absorbance, A.

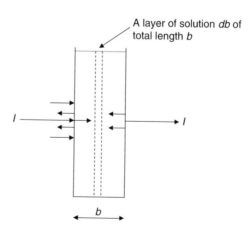

A layer of solution db of total length b

FIGURE 9. *Attenuation of radiation by a cuvette containing a solution.*

Expressed in the usual form, the Beer–Lambert law is, for a single wavelength and a single component:

$$A_\lambda = \varepsilon_\lambda bM \qquad (1.6)$$

Hence, for a given wavelength and a single component, absorbance is a linear function of concentration of that component. However, this equation is based on a number of assumptions, including:

- The radiation is perfectly monochromatic.
- There are no uncompensated losses due to scattering or reflection.
- The radiation beam strikes the cuvette at normal incidence.
- There are no molecular interactions between the absorber and other molecules in solution.
- The temperature remains constant.

These assumptions are not always met and can cause deviations from ideal Beer–Lambert law behaviour.

2.4. Nomenclature

The literature is full of conflicting and often confusing terminology. This book follows the accepted terminology given in Table 2, which is a compilation of terms found in the literature [6] and the IUPAC 'Orange Book' [7], and indicates the alternatives used.

TABLE 2. *Compilation of spectrophotometric nomenclature*

Accepted		Meaning	Alternatives	
Symbol	Name		Symbol	Name
T	Transmittance	$\dfrac{I}{I_0}$	τ (IUPAC)	Transmission factor Transmittancy
A	Absorbance Internal absorbance (IUPAC)	$\log_{10}\dfrac{I}{I_0}$	OD, D, E	Optical density Extinction
$A_{1\,cm}^{1\%}$	Extinction value	$\dfrac{10\varepsilon}{M}$	$E_{1\,cm}^{1\%}$	
A	Absorptivity (c is concentration in g L^{-1})	$\dfrac{A}{bc}$	k	Extinction coefficient, Absorbancy Index
ε	Molar absorptivity Molar absorption coefficient	$\dfrac{A}{bM}$	a^M	Molar extinction coefficient
B	Path length Absorption path length (IUPAC)	Path length cm	l or d	
M	Concentration	mol L^{-1}	c (IUPAC)	

3. FACTORS AFFECTING THE QUALITY OF SPECTRAL DATA

Fascinating though they are, the primary concern of this book is not the theoretical aspects of UV spectrophotometry of materials in aqueous solution. It is concerned with the practical application of the technique and the production of reliable spectral data of known quality. Some of the requirements for solution spectra have been noted previously.

3.1. Good spectroscopic practice [8]

Good spectroscopic practice is a set of pragmatic practical actions and operations that assist in ensuring accurate and reliable measurements for solution spectrophotometry.

The following is a list of some of the more important steps that have to be ensured.

1. The spectrometer is in a proper state of calibration and is well maintained at all times.
2. The solution concentration is as free as possible from weighing, volumetric and temperature errors.
3. The compound to be examined is completely dissolved; ultrasonic treatment as routine is highly recommended.
4. The solution is not turbid—filter if necessary—and that there are no air bubbles on the cuvette windows.
5. Adsorption on the windows is not occurring.
6. The cuvettes are clean and oriented consistently in the light beam.
7. The reference solvent is subject to *exactly* the same procedures as the sample solution.
8. The spectral bandwidth of the spectrometer is correct for the expected natural bandwidth if absorbance accuracy is important.
9. Important regions of the spectrum are measured with the sample absorbance lying between 0.8 and 1.5 A if absorbance accuracy is important. Adjust the cuvette length rather than concentration, if possible.
10. Stray-light is not responsible for negative deviations from the Beer–Lambert law at high absorbance, particularly if the solvent absorbs significantly.
11. Regular checks of absorbance and wavelength accuracy are carried out, and it is checked if the stray-light is within specification.
12. The instrument manufacturer's recommendations are observed.
13. The environment of the instrument is clean and free from external interference. Particular attention should be paid to electrical interference, thermal variations and sunlight.
14. All persons operating the spectrometer and/or preparing samples are properly trained in following the requisite procedures and practices.

3.2. Instrumental criteria

The role of the spectrometer in providing the integrity of data is fundamental to the end result. If the analytical practitioner cannot have faith in the reliability of the basic analytical signal within predetermined limits, then the information generated will be utterly useless.

Knowledge:
Derived from combining valid
information sources

Valid Sample Information:
Derived from good data using
validated application software
which has verified algorithms

Good Spectral Data:
Derived from relevant samples
using validated methods developed
on qualified equipment/validated
computerised analytical systems

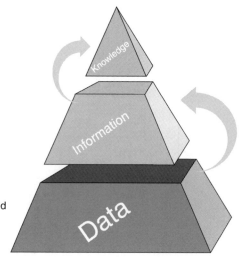

FIGURE 10. *Quality information and knowledge based on good spectroscopic practice.*

Reliability of the data quality should be linked to performance standards for the spectrometer, in addition to having a regular maintenance programme. Quality must be built into analytical procedures, based on the firm foundation of good measurement data and sample practices enshrined in good spectroscopic practice. This process is illustrated in Fig. 10.

3.2.1. Calibration and qualification of the spectrophotometer

The key factors involved in ensuring good spectroscopic data from a UV spectrophotometer for solution measurements are wavelength accuracy and reproducibility, absorbance accuracy and reproducibility and stray-light. The resolution of the spectrometer is of importance only if the chromophores contain absorption bands that are relatively sharp. For most purposes, a spectral bandwidth of 1–2 nm will be suitable.

3.2.2. Wavelength accuracy and reproducibility

For most routine purposes, a solution of holmium oxide in perchloric acid will provide a convenient method for routinely checking the calibration of the wavelength scale. Figure 11 shows a typical spectrum. The values are known to within ±0.2 nm and are adequate for most solution work. If wavelengths in the region below 241 nm are needed, then either atomic line sources such as a vapour discharge lamp or other rare earth solutions may be used.

3.2.3. Absorbance accuracy and reproducibility

For most routine purposes, a solution of potassium dichromate in sulphuric acid will provide a convenient method of routinely checking the calibration of the absorbance

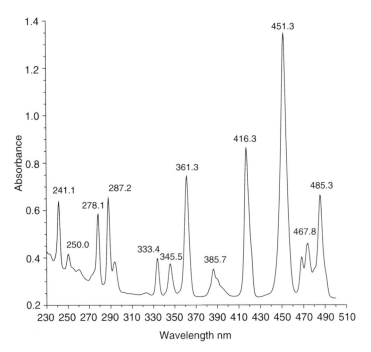

<figure>FIGURE 11. *Holmium perchlorate solution, 10 mm path length and 1 nm SBW. For the latest values J C Travis et al, J. Phys. Chem. Ref. Data, 34(1) (2005) 41–56.*</figure>

scale at four wavelengths in the UV, at 235, 257, 313 and 350 nm. Figure 12 shows a typical spectrum with the $A_{1cm}^{1\%}$ values plotted as a function of wavelength. Certified solutions containing perchloric acid are now widely available in sealed quartz cuvettes.

3.2.4. Stray-light

Stray-light causes deviations from the Beer–Lambert law and limits the upper limit of the absorbance scale. Figure 13 illustrates the effect.

The cutoff filter method is satisfactory for the majority of routine applications. It must always be borne in mind that the observed instrumental stray-light (ISL), is a function of the sample: the measurement of x percent ISL with a cutoff filter does not mean that x percent will again be present when a different absorber is in the beam. It is better to regard the filter method as one that detects stray-light rather than measures it.

The solutions and liquids listed below are recommended as standard cutoff filters. They are also the recommendations of the American Society for Testing and Materials (ASTM) and are generally accepted as industrial standards (Table 3). Compared to glass filters, they have the advantages of reproducibility and freedom from fluorescence. It should be noted that the cells used must be clean, free from fluorescence and with as high a transmission as possible in the region under investigation. Attention to these factors is particularly important when measuring stray-light below 220 nm.

The absorbance below 190 nm is strongly affected by dissolved oxygen. Pure nitrogen should be bubbled through for several minutes before use, and the water should be freshly

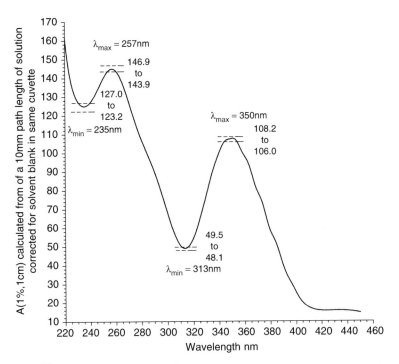

FIGURE 12. $A^{1\%}_{1\,cm}$ values for a 60 mg L^{-1} solution of potassium dichromate in dilute sulphuric acid at 25° C.

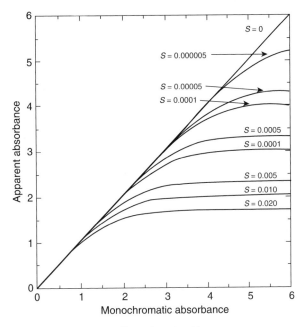

FIGURE 13. *Effect of stray-light on the Beer–Lambert law.*

TABLE 3. *Cutoff filters for stray-light tests. Path length 10 mm*

Spectral range (nm)	Liquid or solution
170–183.5	Water
175–200	Aqueous KCl (12 g L^{-1})
195–223	Aqueous NaBr (10 g L^{-1})
210–259	Aqueous NaI (10 g L^{-1})
250–320	Acetone
300–385	Aqueous $NaNO_2$ (50 g L^{-1})

distilled. Water purified by ion exchange methods may contain significant amounts of organic impurities. Different concentrations or path lengths may be used to displace the absorption edge of these filters so that they can be used in other regions. The absorbance of potassium chloride solution increases significantly with temperature by about 2 percent per °C.

For most purposes, the apparent absorbance of these filters should be more than 2, giving an ISL value of <1%.

3.2.5. Resolution

The best performance of a spectrometer will only be attained, in terms of both absorbance and wavelength accuracy, if careful consideration is given to the resolution of the monochromator. Since resolution is a function of slit width as well as dispersion of the instrument, the choice of slit setting is a critical one. Most modern instruments use grating monochromators that provide constant dispersion with wavelength. The smaller the spectral bandwidth, the greater the resolution, but the corresponding reduction in energy means that the signal-to-noise ratio falls. It is therefore necessary to select the smallest possible slit width that gives an acceptable noise level. When measuring an absorbance band in a high-resolution instrument, it is recommended that the spectral bandwidth (SBW) should not exceed 10 percent of NBW of the band. There is a simple check for the resolution of an instrument.

Record the spectrum of a 0.02% v/v solution of toluene in hexane compared with a solvent blank. The ratio of the maximum at 269 nm and the minimum at 266 nm gives a measure of the resolution of the instrument. A set of spectra is shown in Fig. 14 and the observed ratios in Table 4. The ratio values are within ±0.1 for temperatures between 15 and 30°C, and concentrations of toluene between 0.005 and 0.04% v/v.

TABLE 4. *Observed ratios and spectral bandwidth for 0.02% v/v toluene in hexane*

Spectral bandwidth (nm)	Observed ratio
0.25	2.3
0.5	2.2
1.0	2.0
2.0	1.4
3.0	1.1
4.0	1.0

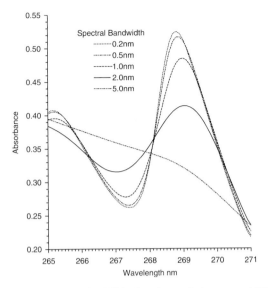

FIGURE 14. *Variation of spectrum of 0.02% v/v toluene in hexane at 25°C with spectral bandwidth.*

3.3. Optimal spectrophotometric range

For many measurements, we have to obtain the most accurate and precise value that we can, given the performance of the instrument. In order to do so, we need to operate in the optimal spectrophotometric range for both accuracy and precision. The majority of instruments actually measure the apparent transmittance, T, of the sample, which is converted to the more useful absorbance, A, by

$$\log \frac{I_0}{I} = \log_{10} \frac{1}{T} = \frac{k'_\lambda bC}{2.303} \tag{1.7}$$

Ideally, the transmittance scale on a linear detector is fixed by a 0% T measurement (dark current measurement on the detector) and a 100% T measurement (total illumination of the detector by I_0). A sample attenuates the I_0 intensity signal, and the sample transmittance and hence the absorbance are obtained. All these individual measurements are subject to noise and drift errors and combine to give an overall measurement standard deviation, σ_T. This standard deviation is related to the relative error of measurement, σ_C/C, and may be obtained by rearranging Equation (1.7) and obtaining the partial derivative. Note that the molarity, M, in Equation (1.5) has been replaced by C the concentration in g L^{-1}. The relative error function [9] is given by Equation (1.8):

$$\frac{\sigma_C}{C} = \frac{0.4343}{\log_{10} T} \frac{\sigma_T}{T} \tag{1.8}$$

Hence, the calculation of the relative error would be straightforward if it were not for the fact that the value of overall measurement standard deviation, σ_T, is not independent

TABLE 5. *Contributions to spectrophotometric precision [Adapted from Reference 11]*

Standard deviation of a measurement σ_T	Source of variability	Relative error function
$\sigma_T = k_1$	Thermal detector, amplifier and dark current noise. [Independent of T]	$\dfrac{\sigma_c}{C} = \dfrac{0.4343}{\log_{10}T}\dfrac{k_1}{T}$
$\sigma_T = k_2\sqrt{T^2 + T}$	Shot noise from the detector	$\dfrac{\sigma_c}{C} = \dfrac{0.4343}{\log_{10}T}k_2\sqrt{1 + \dfrac{1}{T}}$
$\sigma_T = k_3 T$	Cell positioning, non-parallelism errors and incident beam intensity fluctuations	$\dfrac{\sigma_c}{C} = \dfrac{0.4343}{\log_{10}T}k_3$

of the value for T. A detailed theoretical study of the sources and dependencies of the relative error has been made [10]. Ingle *et al.* were able to derive three expressions for σ_T. These are given in Table 5.

If we assign typical values for k_1 and k_2 of $\pm 0.3\%$ T and $\pm 1.3\%$ T for k_3, then we are able to draw graphs of the three functions. This is shown in Fig. 15.

These are derived from theoretical considerations, but similar experimentally determined relative error curves are often observed in practice. The first plot is often found with older single-beam instruments where the error is at a minimum in the 0.4 to 0.6 A range.

Modern double-beam instruments tend to have a broad minimum from 0.6 A up to 1.5 A, depending on the stray-light performance. For diode array spectrophotometers, DAS, generally the minimum tends to be extended to lower absorbance values.

Figure 16 gives an example of experimentally obtained relative error curves at three wavelengths, 240, 426 and 636 nm. Note that the visible wavelengths have a very similar broad minimum from about 0.2 to 2 A, where the relative errors are very small (<0.2%). However, this range is much smaller in the UV region, and, at 240 nm, this range is about 0.2 to 1 A. This range will become even smaller as the wavelength decreases. Particular care has to be taken when working below 220 nm, and accurate measurements are extremely difficult below 200 nm.

Another approach for determining the optimal accurate absorbance range is to use the method of Vandenbelt, Forsyth and Garrett (VFG) [12]. This is a very useful and simple method that is now not so well known. It consists of making a series of solutions over a range of concentrations to cover the absorbance range. From the concentration and the observed absorbance value, the absorbance of a 1% solution in a 1-cm cuvette is calculated, $A_{1\,cm}^{1\%}$. Ideally, all the values will be the same within an experimental uncertainty. However, in practice, this is not observed. At low values of absorbance, $A_{1\,cm}^{1\%}$ values tend to be high, and the converse is also true. The latter effect is usually attributed to stray-light, whereas the former is a failure of Beer's law for reasons that are not clearly understood.

Figure 17 shows an experimental VFG study of a corticosteroid, betamethasone-17-valerate, in ethanolic solution at 25°C using a double-beam spectrophotometer at 238 nm. The accepted value for the $A_{1\,cm}^{1\%}$ of this compound is 325 under the conditions of test. This value is observed if the absorbance range is between 0.6 and 1.9 A. If more concentrated solutions had been made and measured, the $A_{1\,cm}^{1\%}$ values above 2 A would

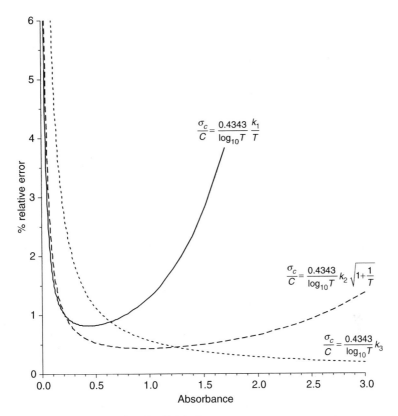

FIGURE 15. *% Relative error plots for Table 5.*

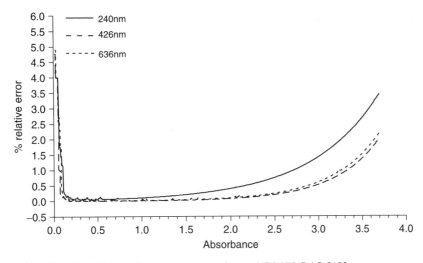

FIGURE 16. *Experimental relative error curves for an HP8450 DAS [13].*

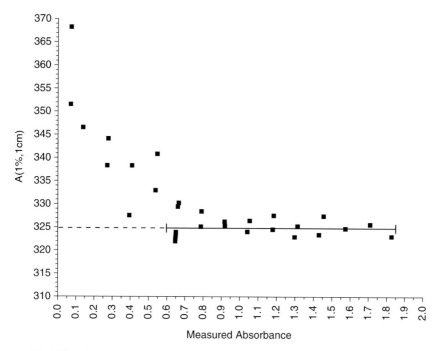

FIGURE 17. *VFG plot for betamethasone-17-valerate in ethanol at 25°C and 238 nm.*

have decreased, making the curve more sigmoidal. Note that $A_{1\,cm}^{1\%}$ values below 0.3 A yield significantly higher values.

The conclusion is that, for the most accuracy, it is desirable to perform relative error or VFG experiments on the spectrometer used under the desired operating conditions. The importance of carrying out this work increases as the wavelengths approach the extremes of the operating range of the spectrophotometer, in particular the ultraviolet.

References

1. R. Keller, J-M. Mermet, M. Otto, M. Valcárcel, H.M. Widmer, *Analytical Chemistry*, 2nd Edition, Wiley–VCH (2004), 741.
2. I. Fleming, D.H. Williams, *Spectroscopic Methods in Organic Chemistry*, McGraw-Hill (1966).
3. C.N.R. Rao, *Ultraviolet and Visible Spectroscopy*, 2nd Edition, Butterworths (1967).
4. R.E. Dodd, *Chemical Spectroscopy*, Elsevier, Amsterdam (1962).
5. H.H. Jaffé, M. Orchin, *Theory and Applications of Ultraviolet Spectroscopy*, John Wiley (1962).
6. *Analytical Chemistry*, 46 (1973), 2449.
7. IUPAC Compendium of Analytical Nomenclature, Definitive Rules, 3rd Edition, Blackwell Science (1997).
8. C. Burgess, T Frost, *Standards and Best Practices in Absorption Spectrometry*, Blackwell Science (1999), 15.
9. D.A. Skoog, D.M. West, F.J. Holler, *Fundamentals of Analytical Chemistry*, 7th Edition (1996), Chapter 24A.

10. L.D. Rothman, S.R. Crouch, J.D. Ingle Jr., *Analytical Chemistry*, 47 (8) (1975) 1126.
11. D.A. Skoog, D.M. West, F.J. Holler, *Fundamentals of Analytical Chemistry*, 7th Edition (1996), Chapter 24A, Table 24-4.
12. J.M. Vandenbelt, J. Forsyth, A. Garrett, *Industrial and Engineering Chemistry*, Analytical Edition 17 (4) (1945) 235.
13. C. Burgess, *Advances in Standards and Methodology in Spectrophotometry*, in C. Burgess, K.D. Mielenz (Eds.), Elsevier, Amsterdam (1987), 307.

UV-Visible Spectrophotometry of Water and Wastewater
O. Thomas and C. Burgess (Eds.)
© 2007 Elsevier. All rights reserved.

CHAPTER 2

From Spectra to Qualitative and Quantitative Results

O. Thomas[a], V. Cerda[b]

[a]Université de Sherbrooke, 2500 boulevard de l'université, Sherbrooke, J1K 2R1,
Québec, Canada; [b]Universitat de les Illes Balears, Carretera de Valldemossa, Km 7.5, 07122,
Palma de Mallorca, Spain

1. INTRODUCTION

The main objective of this chapter is to present the different methods for the exploitation of UV-visible absorption spectra of samples. If quantitative methods, based on the use of the Beer–Lambert law, are well known (simple absorptiometry, multicomponents procedures, etc.), the qualitative exploitation of data is less employed. However, this missing step in spectrophotometric measurement is more often very useful for an understanding of the relationship between sample spectra or even for the "mathematical pretreatment" of the signal. Curiously, this approach is obvious for IR spectra, with the Fourier transformation of the signal, for example, but rarely envisaged for UV-visible applications. Different methods, depending on the number of constituents to be determined and on the complexity of sample, are presented. Before using one of the following methods, it is assumed that no saturation occurs when the spectrum is acquired, in order to be sure that the additive property of the Beer-Lambert law is applicable. For the purpose a good spectroscopic practice as described in Chapter 1 shall be followed.

In the first part, some basic tests or transformation methods can be proposed for qualitative exploitation of spectra. If only one spectrum is available, the user can extract some absorbance values at given wavelengths, calculate the derivative signals (the second one is of particular interest for peak or shoulder identification) or estimate a "shape factor", this new parameter being useful for the treatability study of wastewater (explained in Chapter 10). Where two spectra are concerned, the most evident step is a comparison between them (their shape) either direct or after normalisation, in a given window of wavelengths. This step must not overshadow the great interest of the arithmetical handling of spectra and particularly a simple differentiation. Moreover, while studying a set of spectra, the research of isosbestic point(s), direct or hidden (revealed after normalisation), is very important in order to check an eventual conservation of composition, either quantitative or qualitative.

Then, the different methods available for quantitative analysis will be reviewed. The problem of determining one or two or more components can be solved with the usual methods based on the absorbance measurement at one or several wavelengths, if the optical response of the solution is free of interferences. Unfortunately, for water and wastewater analysis, there is always either physical (diffuse absorption of particles, for example) or chemical interference (e.g. overlapping peaks due to competitive absorbance of compounds), so that more robust methods have to be chosen. The simpler of them are probably the derivative techniques, because they offer to the user the possibility to check the sample quality in a more robust way. Multicomponent procedures based on statistical assumptions and algorithms are then explained. Special interest is shown for a semi-deterministic approach designed for the study of water and wastewater quality. The section concludes with some kinetic consideration and tools, very useful for the study of systems evolution.

2. BASIC HANDLING OF UV SPECTRA

The different qualitative methods are presented in Fig. 1. All these procedures are easy to carry out either directly with the control software of the spectrophotometer, or with the use of any calculation commercial software (for example, Microsoft Excel). They are classified with respect to the number of spectra to be considered and thus to the knowledge level to be reached [1]. Pure statistical procedures (for example, Factor analysis) are not presented in this part. On one hand, computation is more or less complex, and on the other hand, the user is diverted from the meaning of the raw data and from the construction of its own experience.

2.1. One spectrum transformation

Before considering more advanced qualitative tests, i.e. those based on the use of the derivative signals, it can be interesting to present a simple transformation leading to the "visualisation" of the UV response.

2.1.1. Coloured scale

The simplest test is based on the exploitation of the UV spectrum at three wavelengths chosen for their significance [2]: 210 nm corresponding to the presence of nitrate (see Chapter 6), 240 nm allowing the discrimination between soluble organic matrix and suspended solids and 320 nm for suspended solids only. According to the absorbance values, a number and a colour can be proposed in order to give a simple classification and an estimation of the main parameters (Table 1). Notice that this (very) simple classification takes into account the evolution of organic pollution (biodegradation and nitrification).

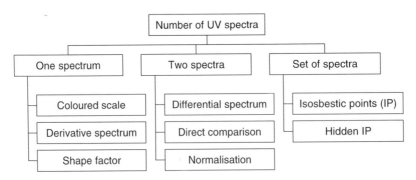

FIGURE 1. *Qualitative methods for UV-visible spectra handling.*

TABLE 1. *Coloured scale for water and wastewater quality, based on absorbance values at 210 nm (λ1), 240 nm (λ2) and 320 nm (λ3)*

Absorbance values	Code	Coloured scale	BOD5 (mg/L)	COD (mg/L)	TSS (mg/L)	Nitrates (mg/L)	Example
λ1 <0.5 λ2 <0.2 λ3 <0.05	1	Blue	<10	<20	<20	<1	Natural water without organic matter nor nitrate or treated and denitrified wastewater
λ1 >0.5 λ2 <0.2 λ3 <0.05	2	Green	<10	<20	<20	>10	Natural water with nitrate or efficient biological wastewater treatment plant outlet
λ1 >0.5 λ2 >0.2 λ3 >0.2	3	Yellow	<50	<150	<50	—	Biological wastewater treatment plant outlet
λ1 >1.0 λ2 >0.5 λ3 >0.2	4	Red	>100	>200	>100	—	Raw wastewater

This coloured scale can be compared to codification systems for water quality mapping.

2.1.2. Derivative spectra

A priori, the derivative spectra do not increase the information content of the normal spectra, but they allow analyzing this information from a more interesting point of view. The study of the derivatives gives some information about the slope of the spectrum and shows more clearly their shoulders and inflexion points, i.e. gives better information about its fine structure. This allows a better characterisation of a compound or shows the deformation due to the presence of a foreign compound. The use of derivatives allows removing the undesired contribution of turbid media.

Figure 2 shows the three successive derivatives of a normal spectrum of uric acid solution. The first derivative corresponds to the slope of the normal spectrum, with maxima corresponding to the increase of the absorbance with wavelength, and minima appearing after the maxima of the usual spectrum. On the other hand, the curve goes through a zero value, which corresponds to the maxima of the usual spectrum.

The second derivative spectrum corresponds to the slope of the first derivative spectrum. In this case, there are several points where the curve has a zero value and minima corresponding to the maxima of the direct spectrum. The third derivative spectrum does not bring supplementary relevant information.

Some generalisations can be drawn from these observations:

- The spectra become more complex with the increase of the derivative degree. The derivative curve has as many occurrences of zero as its derivative degree, having $n + 1$ bands alternating between positive and negative.

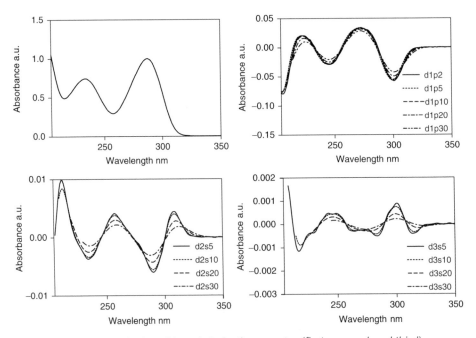

FIGURE 2. *Spectrum of uric acid, and derivative spectra (first, second and third) calculated with several differentiation steps (2, 5, 10, 20 and 30 nm). For example: d2s10 is the second derivative spectrum for a diffferentiation step of 10 nm.*

- Among all these bands, those at the centre are more intense.
- The maximum of the usual spectra corresponds to a zero for the odd derivative spectra, and to a maximum or a minimum for the even derivatives.
- The size (intensity) of the bands decrease with the derivative degree (note the ordinate values of the scales in the graphs of Fig. 2)
- The intensity of the bands strongly depends on the original bandwidth of the direct spectrum. Derivatives make discrimination that benefits the narrower bands.

Taking into account the previous considerations, the second derivative can be considered a good compromise between the significance of the resulting spectra and the signal-to-noise ratio. Actually, the derivative computation is more a differentiation than a true mathematical calculation, since spectrophotometer control software often propose the choice of a differentiation step with a given derivatisation of the initial spectrum. This choice has to be carefully made before further treatment of the UV signals, particularly for the second and eventually third derivative calculations (Fig. 2). A general recommendation can be proposed with a step value close to the quarter of the peak width, which corresponds approximately to 10 nm.

An interesting point can be to understand the practical significance of the second derivative. First, peaks and shoulders can easily be located (corresponding to a minimal value) as well as inflexion points (corresponding to zero). Second, the second derivative value is related to the peak shape and height. Let us consider the estimation of the first "derivative":

$$\frac{dA_{\lambda n}}{d\lambda n} = \frac{A_{\lambda n} - A_{\lambda(n-h)}}{h} \qquad (2.1)$$

where h is the differentiation step, and $A_{\lambda n}$ and $A_{\lambda(n-h)}$ are the absorbance values at the wavelengths λn and $\lambda(n - h)$, respectively.

For the second derivative, the relation is:

$$\frac{d^2 A_{\lambda n}}{d\lambda n^2} = \frac{\left(dA_{\lambda(n+h)}/d\lambda(n+h)\right) - \left(dA_{\lambda n}/d\lambda n\right)}{h} = \frac{A_{\lambda(n-h)} + A_{\lambda(n+h)} - 2 * A_{\lambda n}}{h^2} \qquad (2.2)$$

If we consider that, for a peak (Fig. 3), the maximum of absorbance Am corresponds to $A\lambda n$, and $A*m$ is the absorbance value of the middle of the segment $A\lambda(n-h)-A\lambda(n+h)$, the previous relation becomes:

$$\frac{d^2 A_{\lambda n}}{d\lambda n^2} = -2 \cdot (Am - A^* m)/h^2 \qquad (2.3)$$

A last point is the limit of the use of the second derivative for real samples with interferences (suspended solids, other solutes, etc.). The basic assumption is that the derivative value resulting from interferences is close to zero inside the peak or shoulder wavelength range. This is possible when the corresponding spectra part can be considered to be linear (Fig. 4). The only case where the second derivative can be used for phenol estimation is for sample 2, around 290 nm where the initial spectra is linear (and close to zero). Notice that in the 240-nm region, the interferences influence (very strong on initial spectra) leads to the impossibility of considering the second derivative for use other than qualitative information.

O. Thomas, V. Cerda

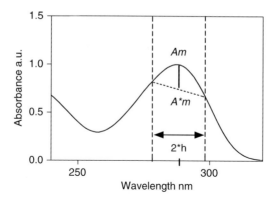

FIGURE 3. *Significance of the second derivative signal. Example of uric acid: peak second derivative (288 nm) = −0.005 (for a differentiation step of 10 nm), compared to Am − A*m = 0.25.*

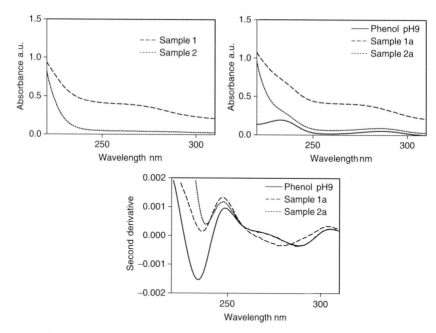

FIGURE 4. *Standard addition of phenol to wastewater – spectra and second derivatives (raw samples: 1 and 2, samples with phenol: 1a and 2a).*

2.1.3. Shape factor

The characterisation of spectra structure is very important as the existence of peaks or shoulders is related to the presence of one or several absorbing compounds. In order to quantify this property, the ratio, for a peak wavelength, for example, between the value of the second derivative and the corresponding absorbance is considered to be of interest [3]. The choice of the use of the second derivative is obvious and known to be a good compromise to reveal any signal perturbation [4,5]. The absorbance value weighting minimises an eventual concentration effect.

A shape factor (SF) is thus defined, at each wavelength corresponding to a peak or shoulder, by:

$$SF = -\frac{D(\lambda)}{A(\lambda)} * H * 10^2 \qquad (2.4)$$

where $D(\lambda)$ is the value of the second derivative measured at the wavelength λ, $A(\lambda)$ is the absorbance value measured at the same wavelength λ and H, the width at the half height of the peak (difference of wavelengths calculated from the second derivative spectrum; see Fig. 4). Considering the second derivative value and sign (negative for a peak), the initial ratio is transformed $(*(-100))$.

According to the value of the SF, UV spectra can be classified into three groups (Fig. 5):

- The first group, $SF > 4$, is composed of samples with UV spectra showing specific absorption peaks or shoulders revealing the presence of major components. In this case, the comparison of UV spectrum of a sample to a UV spectra library may allow the identification of pollutant.
- The second group, $0.1 < SF < 4$, includes samples giving monotonous UV spectra. In this case, there exists a large probability to consider the studied effluents as complex

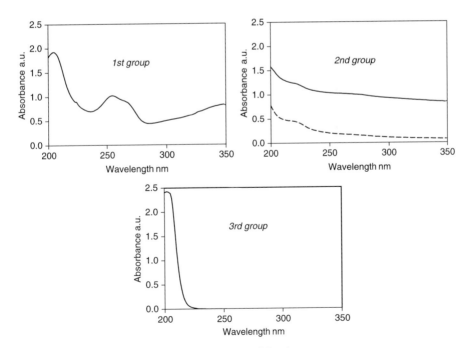

FIGURE 5. *Typology of spectra according to the SF value.*

mixtures or due to the presence of colloids or suspended solids, and more complete investigations must be carried out for a better discrimination.
- The last group, SF < 0.1, corresponds to nonabsorbing samples or mineral effluents. As said before, this phenomenon is actually rare for organic wastewater.

2.2. Two-spectra comparison

More than spectrum characterisation, a comparison of two spectra can often be of great interest.

2.2.1. Differential spectrum

This operation is very simple and useful, and can be performed with most laboratory spectrophotometers. By subtracting a spectrum by another (i.e. the difference of absorbance values for each wavelength), the result can give relevant information as, for example, for a tentative of interferences elimination, if possible, or for the study of a treatment effect, as filtration. In Fig. 6, the difference 1 corresponds to the efficiency of a physical process with suspended solids removal. The same result could be achieved with a filtration step, the difference between raw and filtered samples giving the same diffusion shape. In this case, the difference between the spectra of the raw sample and the filtrate may give some information on the solids retained on the filter. The limit of the method is shown in the same figure, when absorbing compounds appear (like nitrate). The difference between inlet and outlet leads to negative values of absorbance, which is physically impossible. For the treatment plant 2, the nitrification step is efficient. Thus, it can be proposed to use differential spectra when a simple evolution of a sample is studied.

The previous examples deal with the spectra of real samples. Another way to proceed is to subtract, from the real spectrum, a spectral contribution corresponding to the presence of a given absorbing compound. For example, in natural water, the presence of nitrate can hide the optical response of organic compounds, even if the water has been filtred with a very low cut-off membrane (Fig. 7). The determination of nitrate concentration (by conventional analytical methods or by UV method; see Chapter 5) allows to subtract the part of the spectrum related to nitrates from the initial spectrum.

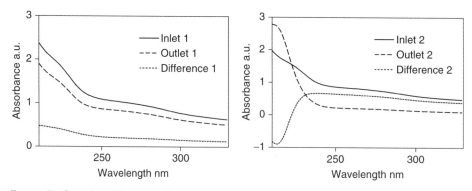

FIGURE 6. Spectra differences for wastewater treatment plants samples (inlet and outlet).

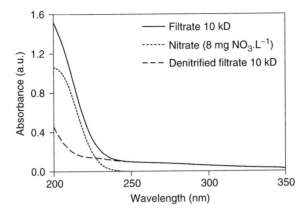

FIGURE 7. *Revelation of organic matrix by spectra difference.*

This basic handling makes possible the revelation of the UV response of the organic matrix. The presence of carboxylic acids at low concentration can explain the "denitrified filtrate" spectrum shape.

2.2.2. *Direct comparison*

This data exploitation is less used but interesting if a comparison between spectra is required, in order to check the general quality and origin of wastewater, for example. For each wavelength, the absorbance value of one spectrum is plotted against that of the second spectrum. The initial spectra shape is lost, but the obtained graphical relation is relevant (Fig. 8).

Considering the result, a straight line or a curve, the studied spectra can have the same shape (homothetic) or be very different. Depending of the curve type, information

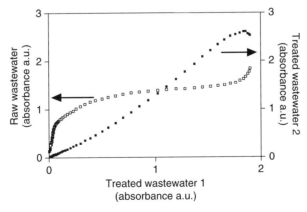

FIGURE 8. *Direct comparison of absorbance values (the treated wastewater is considered the reference type).*

on sample nature can be given [6]. Notice that the higher values of absorbance must often not be considered because of saturation risk, and that the slope of a straight line leads to the homothetic ratio.

2.2.3. Normalisation

In several cases, spectra normalisation is a preliminary step to a further study [1]. This operation leads to give a same area (arbitrarily chosen) under the studied spectra. Considering that UV spectra of aqueous samples are often related to dilution phenomena, the normalisation step tries to prevent this effect. In practice, the area of spectra is given by the sum of absorbances between two given wavelengths (e.g. 200 and 350 nm). More generally, if the absorbance values are acquired every h nanometres, the area can be calculated by:

$$Area = \sum_{\lambda=200}^{350} A(\lambda) \times h \tag{2.5}$$

For a chosen norm, the corrected absorbance values $A^*(\lambda)$ must be calculated from the product of the measured absorbance values of each spectrum by the ratio of the value of the norm divided by the area of the spectrum.

$$A^*(\lambda) = A(\lambda) \cdot \frac{Norm}{Area} \tag{2.6}$$

In the previous case (direct comparison of samples), two normalised spectra of the same type would lead to a straight line with a slope equal to 1. Their corresponding normalised spectra are thus superposed (Fig. 9).

In the general case, where the shape of spectra varies from one spectrum to another, the normalisation step leads to crossing spectra, as shown in Fig. 10.

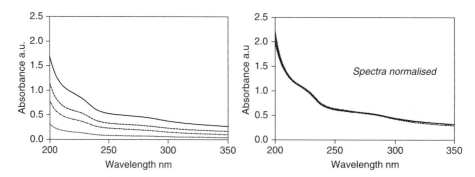

FIGURE 9. *Normalisation of spectra of same shape (raw wastewater).*

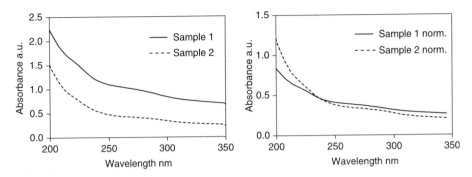

FIGURE 10. *Normalisation of spectra of different shapes.*

2.3. Evolution study from a spectra set

The simplest method to study a spectra set evolution is to display the spectra on Fig. 10. Several examples can be further shown for degradation tests. More than the general evolution, leading in this case to the decrease of spectra shape and thus of absorbance values, isosbestic points can appears with time.

2.3.1. Isosbestic points

If all spectra, or at least several, cross together at one point, this particular point is called isosbestic point (IP). Several isosbestic points may exist from a set of spectra. One classic example of such isosbestic points is, for instance, a set of spectra of one component in solution, at various pH values, which shows an equilibrium between acidic and basic forms, the proportion of these depending on pH value. Most of the studies on isosbestic points were made on reacting systems involving pure components artificially mixed in the laboratory (for example, see Reference 7); this phenomenon can be observed for the discharge of wastewater into a river (Fig. 11).

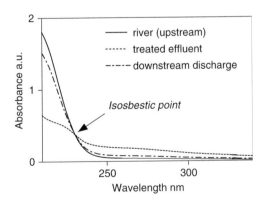

FIGURE 11. *Study of treated wastewater discharge into a river showing an isosbestic point.*

In practice, because of some instrumental or environmental errors, an isosbestic point is rather a small surface than a real point, and a procedure has been proposed for its detection [8,9]. An isosbestic point is defined by the wavelength λ_{IP} (or λ_{IP*} for hidden IP) as a point where the apparent coefficient of variation (CV^*) is lower than a limit value (fixed to 2.5%, value obtained from a statistical study on repeatability):

$$CV^* = \frac{\sigma^*(\lambda)}{\overline{A(\lambda)}} \times 100 \qquad (2.7)$$

where CV^* is the apparent coefficient of variation (%), $\overline{A(\lambda)}$ is the average of absorbance values at the wavelength λ (a.u.) and $\sigma^*(\lambda)$, is the standard deviation estimation of absorbance values at the wavelength λ (a.u.).

The search for points called outliers, responsible for a coefficient of variation greater than the fixed value is based on a statistical test (Dixon test). The UV spectra eliminated following this test are considered as not representative of the studied flux. Then, a final statistical test is carried out (Rank test, for example) in order to check if the revealed point is a true isosbestic point. This final test is carried out at $\lambda_{IP} \pm 10$ nm.

The presence of at least one isosbestic point shows that there is a mass conservation between all samples, which can thus be considered as a mixture of compounds. More often, this indicates the presence of only two major mixtures (considered like pure compounds), the concentrations of which are linked in a way that the mass balance is conserved [10]. More precisely, there is a fixed linear relation between the concentrations of the two components (or mixtures of fixed mass composition) of the form:

$$a_1 C_{1i} + a_2 C_{2i} = 1, \quad \forall i \qquad (2.8)$$

where C_{1i} and C_{2i} are the concentrations of components 1 and 2 in the mixture i, and a_1 and a_2 do not depend on the mixture i.

The presence of an isosbestic point in Fig. 11 confirms the mass conservation between two mixtures characterised by the presence of anthropogenic organic matter (for treated effluent) and by nitrates (for river), the proportion of which varies according to the sampling place (river, effluent or mixture between discharge and river).

2.3.2. Hidden isosbestic points

More often, no isosbestic point appears in a set of spectra of real samples. This can be explained by several factors as the occurrence of dilution or other physicochemical factors (sedimentation, precipitation, oxidation, etc.). However, in case of quality conservation (simple dilution, for example), a normalisation step can lead to the revelation of at least one isosbestic point in the resulting set of spectra. This isosbestic point is called hidden isosbestic point (HIP).

If a normalisation step is necessary before revealing one hidden isosbestic point, no mass conservation can be assumed, but the effluent quality remains constant [11,12]. In this case, the samples composition is characterised by the presence of the same components but in variable proportions. The global concentration is also variable. Indeed, the normalisation step consists of making artificial conditions of mass conservation.

Figure 12 presents a set of spectra of wastewater sampled at different hours of a day, before and after normalisation.

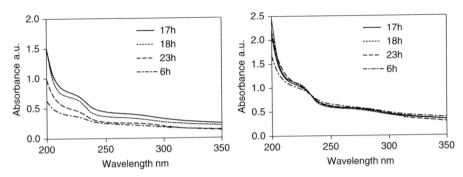

FIGURE 12. *Raw (left) and normalised (right) spectra of wastewater showing hidden isosbestic point.*

The presence of an HIP proves the quality conservation of water. In this case, the samples (wastewater) are assumed to be made of a mixture of two complex components, particles greater than 1.2 μm (total suspended solids, see Chapter 6) and matter smaller than 1.2 μm ("soluble" organic matter, see Chapter 4), the proportion of which vary according to time and weather [13]. The absence of direct IP can be explained by the infiltration of clear parasite water responsible of dilution, especially during night.

This method is very interesting for the study of the qualitative variability of water and wastewater.

2.3.3. *Application: variability estimation*

The variability is defined from the calculation of the ratio of the number of spectra crossing at the isosbestic point (Npi), divided by the total number of spectra (Nt) of the initial set [9]:

$$V = \left(1 - \frac{Npi}{Nt}\right) \times 100 \tag{2.9}$$

The variability is estimated without any knowledge of the medium composition. An application is presented for industrial wastewater in Chapter 9.

3. CONCENTRATION CALCULATION

The main interest in using UV-visible spectrophotometry is for analytical purposes. Historically, a lot of analytical methods are based on the use or colorimetry, with a specific reagent and the visual or photometric detection of the colour of the final solution. The analytical performances have been improved with the final spectrophotometric detection, which is more sensitive and more accurate, with the user being able to check the working wavelength. All these methods, employed for the analysis of a single compound in solution, are based on the Beer–Lambert law. From the spectrophotometric analysis of one compound to the simultaneous determination of specific or global parameters by using

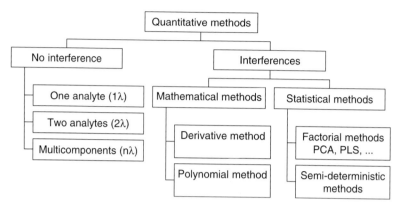

FIGURE 13. *Classification of quantitative methods for UV-visible spectra exploitation.*

matricial procedures, this part reviews the main methods available for water and waste-water analysis. Figure 13 presents the different methods depending on the existence of interferences.

3.1. Ideal case: pure solution with no interference

3.1.1. Simple absorptiometry for one analyte

The Beer–Lambert law (presented in Chapter 1) has some limitations, and the main one is that it is only true for low analyte concentrations. The global relation used for the calculation of the concentration, C, from the absorbance value, A, at a given wavelength, the choice of which is explained hereafter, is:

$$C = f(A) \qquad (2.10)$$

For higher concentrations, a number of corrective factors have to be introduced, as for example the refraction index variation related to the concentration of the analyte. One way to correct this effect is to substitute the value of ε_λ of the Beer–Lambert equation by $\varepsilon_\lambda n/(n^2 + 2)^2$, where n is the value of the refraction index. Usually, this correction is not very high for concentrations lower than 0.01 M. Another effect that can distort the Beer–Lambert law linearity may be the use of a polychromatic radiation. This problem, encountered with instruments using filters (photometers), obliges the use of instruments allowing the selection of narrower wavelength ranges by means of monochromators (spectrophotometers), and are, therefore, more expensive.

Usually, the absorbance measurements are taken in the maximum of the spectrum (peak for example) due to several reasons. On one hand, the maximum sensitivity (greater slope in the calibration curve) is obtained at this wavelength. On the other hand, the centre of the maximum is where the absorbance gradient is minimum vs. the wavelength, which means lower probability of deviations from the Beer–Lambert law due to the polychromatism of the selected radiation. Finally, it will be a lesser variation of the method sensitivity due to the imprecision in positioning the wavelength. Measurements are sometimes performed not at the maximum, but in other places (shoulder, for example), in order to decrease the obtained values and, therefore, do not saturate the instrument response.

In quantitative determinations, usually, a calibration curve is first obtained with the use of several standards. Although a great number of molar absorptivities may be obtained from the literature, a better option is to obtain oneself the calibration curve in exactly the same experimental conditions that will be later applied to the samples. Bibliographic data may, however, be valuable in order to know the sensitivity of the analytical method, determination limits, precision, etc. Several books [14,15] may be cited, in which a great number of tested photometric methods are described for a great number of elements. The molar absorptivity at the analytical wavelength is given for each method, together with the relative standard deviation, path length, interferences, etc.

Sometimes, the matrix of the standards is quite different from the one of the samples to be analyzed. This may give big determination errors (due, for example, to the formation of binary or ternary complexes with other ligands present into the matrix). In this case, it is better to use the standard addition method, where the standards are not independently measured from the samples. In this method, n identical aliquots are taken for each sample, and increasing known quantities of the standard are introduced in them and diluted to the same final volume. Once the graph of absorbance values vs. added concentration is represented, the extrapolation to zero absorbance will give the desired unknown concentration of the analyte.

3.1.2. Two analytes

TWO WAVELENGTHS METHOD   The spectral overlapping of the components of a mixture is one of the most important limitations of the spectrophotometric methods when one component has to be determined in the UV-visible range. When two analytes have to be determined in the same solution, two different wavelengths have to be chosen in such a way that one analyte does not interfere with the other. The general relation between the two concentrations to be determined and the two measured absorbance values is:

$$C1, C2 = f(A1, A2) \tag{2.11}$$

The practical calculation is explained hereafter, and one calibration curve for each analyte has to be drawn for its appropriate wavelength without any other complication.

Another usual and undesired effect is the light scattering produced by the suspended or colloidal particles. This scattering is often considered as nearly constant in the overall UV-visible spectral range. Therefore, an overlapping is produced with the spectrum of the analyte to be determined. In this case, the interference may be easily removed by subtracting to the absorbance values, the absorbance contribution of the particles. This one is measured in a spectral zone where the analytes do not absorb and where only the scattering particles are contributing. If the scattering is not constant, but their slope is, then the derivative spectroscopy may be applied, as will be seen later.

Since it is very often impossible to avoid the spectral overlapping of two analytes, it is necessary to use some mathematical procedure in order to make a discrimination of their spectrophotometric signals.

The classical method is to find the same number of analytical wavelengths as the number of analytes to be determined, trying to select those where the difference between the molar absorptivities of the different compounds is maximum. In this way, the same number of linear equations as analytes are established, and may be solved by means of traditional computing techniques (determinants, for example).

So, if two species give two overlapped spectra, the following equations may be written:

$$A^{\lambda 1} = \varepsilon_1^{\lambda 1} bC_1 + \varepsilon_{\acute{e}}^{\lambda 1} bC_2$$
$$A^{\lambda 2} = \varepsilon_1^{\lambda 2} bC_1 + \varepsilon_{\acute{e}}^{\lambda 2} bC_2 \tag{2.12}$$

where the super indices λi correspond to the lectures made at the two analytical wavelengths i (1 and 2), whereas the subindices correspond to the compounds 1 and 2, respectively.

The molar absorptivity values (ε) of both substances may be obtained from pure standards. Once these values are known, the absorbance values for each sample obtained at both analytical wavelengths allow computing the concentration values from the following equations:

$$C_1 = \frac{A^{\lambda 2} \varepsilon_2^{\lambda 1} - A^{\lambda 1} \varepsilon_2^{\lambda 2}}{\varepsilon_1^{\lambda 2} \varepsilon_2^{\lambda 1} - \varepsilon_1^{\lambda 1} \varepsilon_2^{\lambda 2}}$$

$$C_2 = \frac{A^{\lambda 1} \varepsilon_1^{\lambda 2} - A^{\lambda 2} \varepsilon_1^{\lambda 1}}{\varepsilon_1^{\lambda 2} \varepsilon_2^{\lambda 1} - \varepsilon_1^{\lambda 1} \varepsilon_2^{\lambda 2}} \tag{2.13}$$

This system may only be used if the Beer–Lambert law is applied and there is no mutual influence between the two components.

N WAVELENGTHS METHOD The spectrophotometric resolution of mixtures of two components on the basis of the well-known extended Beer's law relies on absorbance measurements at two different wavelengths and the fulfilment of the law of the additivity of absorbances [16]. This methodology has several shortcomings, such as the use of only two experimental data obtained at two different wavelengths. While the method could in principle be used to resolve up to n components by making measurements at as many wavelengths, it has not been applied to more than two components because its accuracy decreases sharply as the number of involved determinants grows [17]. The precision of the above-described method in the resolution of multicomponent mixtures can be increased by performing measurements at more wavelengths.

Thus, for a given wavelength λ_i,

$$\frac{A_m^{\lambda 1}}{A_{s_1}^{\lambda 1}} = \frac{c_1}{c_{s_1}} + \frac{A_{s_2}^{\lambda_i} c_2}{A_{s_1}^{\lambda_i} c_{s_2}} \tag{2.14}$$

where $A_m^{\lambda 1}$ and $A_{si}^{\lambda 1}$ are the measured absorbances of the sample and of the corresponding standards (S_1 and S_2) at wavelength λ_i C_{S_1}, C_{S_2}, C_1 and C_2 are the concentrations of standards and components of the mixture, respectively.

From the previous relation, it follows that, by plotting the ratio of the measured absorbances of the sample on the standard S_1 against the ratio of the measured absorbances of the standards S_1 and S_2, for different wavelengths, one will obtain a straight line whose intercept and slope will provide the sought C_1 and C_2 values.

Blanco *et al.* [17] compared these two procedures for the determination of binary mixtures with highly overlapped spectra, obtaining better results by the multiwavelength linear regression method.

3.1.3. Multicomponent method by Mutli Linear Regression (MLR)

In theory, if no interference exists in the solution, the generalisation of the additive relation can be applied, providing matrix effects, and any chemical interactions involved are negligible.

$$A(\lambda_i) = \sum_{j=1}^{p} \varepsilon_{j(\lambda i)} Cj + r \qquad (2.15)$$

where *r* is the residual value, i.e. the difference between the measured and calculated values. This value must be minimised by using one of the following methods.

For the whole spectrum, the previous relation must be extended:

$$S = \sum_{i=1}^{n} A(\lambda_i) = \sum_{i=1}^{n}\sum_{j=1}^{p} \varepsilon_{j(\lambda i)} Cj + r(\lambda_i) \qquad (2.16)$$

If the number of measurements (wavelengths, *n*) exceeds the number of components (*p*), then the above system will be over-dimensioned and resolvable by multiple linear regression (MLR). Such a system can be expressed in a matrix form as:

$$L = K \cdot C \qquad (2.17)$$

which entails solving two analytical chemical problems, namely:

First, one must determine the proportionality constants from matrix **K** by using standards of known concentration, and

Once matrix **K** has been determined, one must resolve the unknown mixtures so as to determine the concentration matrix **C** from the following equation

$$\mathbf{C} = (\mathbf{K'K})^{-1}\mathbf{K'L} \qquad (2.18)$$

where **K'** denotes the transpose of **K**.

Matrix **K** can be determined in a number of ways, including the following [18]:

SINGLE OR AVERAGED STANDARDS This is the simplest procedure and involves the use of a single or averaged standard of known concentration for each component and the calculation of the corresponding response factor from:

$$k_i^m = \frac{I_s^m}{c_{s_i}} \qquad (2.19)$$

SEVERAL STANDARDS OF THE COMPONENTS AND THEIR MIXTURES. This option entails calculating the different k_i^m values by regression from standards of different concentrations
of each component or mixtures of known composition. Mathematically, the procedure
involves an equation system for each measuring channel of the form:

$$I_s^m = z^m + \sum_{i-1}^{n} k_i^m c_s \qquad \forall s = 1 \ldots n_s \qquad (2.20)$$

where I_s^m denotes the reading obtained for standards s in measuring channel m, z^m the
independent term of the fitting for each m value, and c_{si} the concentration of component i
in standard s. This equation system can be solved by multiple linear regression, provided
that the number of standards used, n_s is larger than that of components. On solving the
system, one obtains the k_i^m and z^m values, as well as the deviation of the fitting for each
measuring channel, d^m, which can be calculated from:

$$d^m = \sqrt{\frac{\sum_{s-1}^{n_s} \left(I_{s,exp}^m - I_{s,calc}^m\right)^2}{n_s - n}} \qquad (2.21)$$

The d^m values can subsequently be used in resolving the unknown mixture so as
to carry out a weighted fitting of the initial equation in such a way that the measuring
channels with the greatest deviations will have a smaller weight than the rest. Thus, the
equation to be used is:

$$\frac{I^m}{d^m} = k_0^l + \sum_{i-1}^{n} \frac{k_i^m}{d^m} c_i \qquad \forall m = 1 \ldots n_m \qquad (2.22)$$

GENERALISED (MULTIPLE) STANDARD ADDITION METHOD Equation (2.20) can be applied to
a data set obtained by spiking the unknown samples with known amounts of the components to be determined. In this case, the c_{si} values will correspond to the added
concentrations of each component in each standard. As above, the regressions performed for each measuring channel will provide the k_i^m and z^m values, which will
represent estimates of the sensitivity of each component and the sample signal (in the
absence of additive interferences), respectively. Finally, by using Eq. (2.20) with the z^m
values, one can calculate the concentration of each component in the unknown sample.

In practice, MLR is often limited to five variables (components) for the regression,
due to possible collinearity between spectra. However, MLR is the simplest method for
multicomponent analysis because it is easy to understand and use, even for a nonexpert
in matrix calculation.

Multicomponent analysis techniques have opened up new prospects for the resolution
of diverse analytical systems. Frequently, the application of these techniques requires
some chemical or instrumental innovations with respect to the previous procedures.

Among the referred algorithms, the multilinear regression method has often been used
because of its easy implementation, and good results have been obtained in most cases.
The chemical interferents are the main limitation of such techniques, because prior knowledge of each substance that contributes to the overall signal is needed. In this case,
multiplicative interferences can easily be addressed using the multiple standard addition
method. The elimination of additive interferences has not been achieved.

3.2. Real samples: mathematical compensation of interferences

3.2.1. Derivative methods

The interest of derivative signal, and particularly the second one, has already been presented in this chapter (Section 2.1.2).

Different quantitative methods of analysis may be proposed from the use of derivative spectra:

- The derivative value may be used at any wavelength (except in the wavelength where the derivative value is zero), and particularly where the second derivative is minimum (corresponding to the maximum of the normal spectrum).
- In the peak-valley method (Fig. 14), the difference between two maximum–minimum values is used, this method being more sensitive than the previous one.
- In the tangent method, two lines are drawn between two consecutive maxima or minima, and the distance between this tangent and the intermediate maximum or minimum is calculated; this method allows correcting any variation of the base line due to a matrix effect.

The derivative spectra allow knowing with better precision the position of maxima and minima in qualitative analysis. We have seen that an increase in the derivative degree increases the number of peaks in the spectrum, which allows a better characterisation of the substances than the direct spectrum.

By means of derivative spectroscopy, one can appreciate small distortions of the direct spectrum due to the presence of impurities, as well as to discriminate the spectral contribution of overlapping compounds.

Referring to quantitative analysis, derivatives allow discriminating the presence of an overlapping compound in the foot of the peak of a major compound. In Fig. 15, a minor compound appears as the shoulder of a major component. The dotted line corresponds to the minor compound alone, without the interference of the major component.

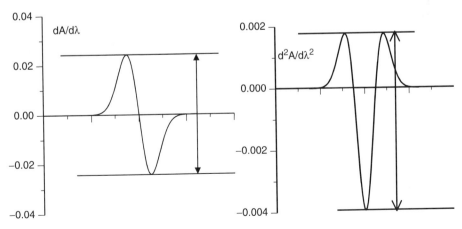

FIGURE 14. *Graphical methods for quantitative analysis.*

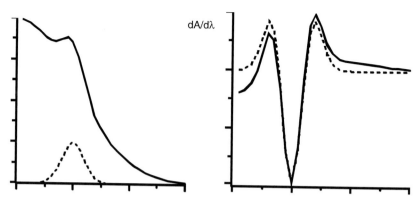

FIGURE 15. *Mixture of the spectra of major and minor components and second derivatives.*

The second derivative spectra are represented, both of the mixture as well as of only the minor compound. As one can appreciate, in these derivative spectra, the contribution of the major compound is strongly reduced.

An important application of derivative spectroscopy lies in the determination of analytes in turbid media. Turbid solutions usually present a continuous increase of the absorbance towards shorter wavelengths and, as a consequence, do not produce any sudden spectral change either in the first or in the second derivative spectrum. Phenol determination in wastewater was one of the first practical applications of the derivative spectroscopy in turbid media [19]. Another study, more recent, deals with the evaluation of the second-derivative determination of nitrate and total nitrogen [20].

When the medium has a wide and nonlinear spectrum, it is often necessary to use a higher derivative degree in order to remove its contribution. However, in order to avoid an important loss of sensitivity, the second derivative is used, together with the tangent method, to remove the background noise.

For bands with similar widths, it is not possible to take advantage of the discriminative power of the derivatives. One can then use the zero crossing method. To the maximum or minimum of each band, there is a correspondence with a zero derivative value; the concentration value of the compounds do not matter. To the inflexion point of the usual spectrum, the second derivative is zero, etc. To these particular points, the value of the derivative is due to the only contribution of the second compound and, in this way, the interference of the first one may be removed. This methodology was used at the very beginning of derivative spectroscopy and was mainly applied to those compounds that had close spectra. One has to note, however, that this methodology requires the use of very reproducible wavelength-positioning spectrophotometers.

3.2.2. Polynomial compensation of interferences

Another way to compensate the interference effects is to modelise their optical response by a simple mathematical function, $f(\lambda)$, allowing to explain the measured absorbance value from the expected response of the p solutes:

$$A(\lambda_i) = \sum_{j=1}^{p} \varepsilon_{j(\lambda_i)} C_j + f(\lambda_i)$$ (2.23)

Several functions can be proposed, but the most adapted seems to be a polynomial one [21].

$$f(\lambda) = \sum_{i=1}^{n} a_i \cdot \lambda^i$$ (2.24)

$$f(\lambda) = a_0 + a_1 \cdot \lambda + a_2 \cdot \lambda^2 + a_3 \cdot \lambda^3 + \cdots + a_n \cdot \lambda^n$$

An interesting aspect of this choice is that, depending of the polynomial degree, several shapes are assumed for the restitution of the interferences effect:

- A zero degree corresponds to a constant shift of absorbance, encountered when the spectrophotometric cell is dirty, for example.
- A first degree is equivalent to a linear response of interferences very often used for the exploitation of chromatographic data (calculation of peak area). The correction procedures of Allen (or Morton and Stub) are based on the same assumption. This solution is equivalent to the use of the second-derivative signal previously described, if we remember that a linear response gives a zero value for the second derivative.
- For higher degrees, any interference response can be fitted with a polynomial response. However, the polynomial degree to be considered must not be too high for preventing the risk of the whole spectrum (interference and analytes) modelisation. This is the reason why a polynomial of third degree is the best compromise [16,22,23]. In some cases, this method can be simplified by neglecting the lower terms and keeping the only term of the third degree for compensation [24].

3.2.3. Other mathematical tools

The following methods are used to increase the signal-to-noise ratio. They contribute to the smoothing of the signal by filtering the response. The use of these methods have been made possible by the significant advances in the field of developing instruments. The diode array and CCD components, on the one hand, have eliminated the mechanical wavelength uncertainty of the spectrophotometers. On the other hand, signal digitalisation and the use of microprocessors have allowed substituting the optical and electronic systems with computational algorithms of derivation, like the Savitzky-Golay method, which simultaneously allows noise removal and signal smoothing.

One of the most important limitations of derivative spectroscopy is the background noise associated with any experimental measurement. This random noise usually has a higher frequency than the signal to be measured. This means that it will give weak but narrow peaks. The use of derivatives will strengthen this kind of peaks in front of the broader and stronger ones of the compounds, which are the most important in the usual spectrum. A solution to this problem is to use the Fourier transform (or indeed better, the fast Fourier transform, FFT), which allows the use of filters in order to remove high-frequency contributions. This solution was not easy to use sometime back, but the introduction of computers as instrumental controllers and, indeed, the development of specialised FFT chips have rendered this alternative very attractive (although not very extended yet). It is enough to use a high-frequency cut-off filter to remove the noise from the signal.

3.3. Real samples: statistical and hyphenated methods

The last group of methods for the exploitation of spectra of real samples involves not only factor analysis procedures and related methods but also the semi-deterministic approach. Both types of methods can be considered as statistical ones, but with a slight difference. Factors analysis and related methods are also described as "black boxes," because they require no information. They function in a manner that is quite opposite to that of MLR methods, for which the response of all compounds of the sample must be known (which is obviously impossible for a real sample). Also, the semi-deterministic approach is based on a "grey box" type of model, where only a part of the information is needed, the other part remaining stochastic.

3.3.1. Factor analysis: PCA, PCR and PLS

The chief limitation of the above-described methods is that they require all the species contributing significantly to the measured signal to be known beforehand. This pitfall has lately been circumvented by developing new chemometric procedures based on advanced multivariate analysis. Starting from the matrix representation of data, these procedures tend to extract the relevant information with the aim of representing the synthesis of results within one or two simple graphs. This qualitative exploitation, leading to the decomposition of spectra, can be associated with a regression step for the estimation of component concentration. It allows the determination of the number of components significantly contributing to the analytical signal, and then permits the spectrum of each individual component to be reconstructed, which finally allows the system to be resolved analytically.

In practice, for spectra exploitation, the main procedure is the principal component analysis (PCA), identifying a set of few factors (the first eigenvectors of the matrix), used for the interpretation of data. Then, any spectrum can be explained as a linear combination of these factors (as a decomposition step), the coefficients of which are the PCA scores.

For the estimation of components concentration, a second step is required, based on a multiple linear regression (MLR, see Section 3.1.3) between the absorbance values and the PCA scores. This can be carried out automatically after the PCA step, with the principal component regression (PCR) procedure (including PCA). This methodology was first applied to analytical chemical problems by Lawton and Sylvestre [25], and has more recently been used in different models by other researchers [26–28]. Finally, the PCA procedure can also be coupled with cluster analysis (CA), as described in a very recent study on the characterisation of industrial wastewater samples [29].

Another method, very often used for NIR analysis, is the partial least squares (PLS) procedure. This method is slightly different from PCR, because the two steps of the last method (decomposition and regression) are carried out at the same time and the decomposition process also includes concentration information. The results (eigenvectors and scores) are thus different and generally more relevant, because they are more related to the concentration data. In fact, both spectral and concentration data are considered and simultaneously decomposed by iteration, and the results depend on the chosen factor number. The PLS method (in fact, two methods, PLS1 and PLS2, are available) is more complex than the PCR method but seems to be more adapted for huge data sets as for spectroscopic application (e.g. NIR for petroleum products).

A comparison between PCR and PLS is difficult because the quality of results is rather close (at least for a small set of data), but some specific advantages of the PLS can be drawn. The intermediate results (eigenvectors) are related to the initial physical data (looks like particular spectra). Moreover, the calibration step is more robust (if the data set is representative), and PLS can thus be used for the study of complex mixtures. Some known drawbacks are the computational time, the need for a large calibration set (representative) and some difficulties in understanding and explaining the resulting model.

These procedures are sometimes present in the built-in software of UV-visible spectrophotometers.

Factor analysis methods allow resolution of multicomponent mixtures when individual contribution of each component is unknown. In broad terms, this methodology yields a solution set for each component whose width depends on the data supplied. Nevertheless, the complexity of the mathematical treatment has actually prevented the resolution of chemical systems with more than three components.

As mentioned previously, one of the main advantages of PLS is that the resulting spectral vectors are directly related to the constituents of interest. This is entirely unlike PCR, where the vectors merely represent the most common spectral variations in the data, completely ignoring their relation to the constituents of interest until the final regression step.

3.3.2. Semi-deterministic method

As with the factor-analysis-based methods, this method too aims to explain any acquired spectrum through a deconvolution step, and to calculate some parameters of interest. It assumes that each spectrum may be considered a linear combination of a p reduced number of particular spectra, which are named "reference spectra" [5]:

$$S_u = \sum_{i=1}^{p} S_{ref_i} + S_{res} \qquad (2.25)$$

where S_u and S_{ref_i} are, respectively, spectra of an unknown sample and of the ith reference spectrum (among p) and the S_{res} residual spectrum is the difference between the acquired and restituted spectra.

These spectra (Fig. 16) are either spectra of specific compounds or of aggregate matrices (residual organics dissolved, colloids, suspended solids). The first group of spectra, very reproducible, is the deterministic part of the model. It includes the compounds that may be found in the type of sample to be examined. The second one, being of experimental or mathematical nature (difference of spectra, for example), can be considered as the stochastic part of the model. Moreover, some of these spectra can be actually related to principal components calculated from the residual matrix. The selection is done between different spectra, which allows taking into account the effect of the main interferences.

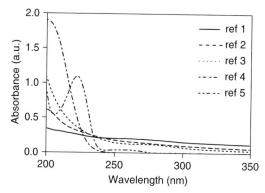

FIGURE 16. *Example of reference spectra, normalised (ref 1: dissolved organics, ref 2: colloids, ref 3: suspended solids, ref 4: nitrates, ref 5: surfactants (DBS)).*

In Fig. 16, the spectrum of suspended solids is the result of the difference between the spectrum of the raw sample and the spectrum of the same sample previously filtered through a 1 micrometer filter. In the same way, the spectrum of the colloidal fraction is the difference between the spectrum of the solution resulting after its filtration through a 1 micrometer filter and that obtained using a 0.025 micrometer filter. Finally, the interferences related to the organic matrix may be represented by the spectrum of the sample previously filtered through a 0.025 micrometer filter. The basis of reference spectra is completed by the spectra of nitrate and surfactant, considered as the deterministic part of the model.

From a mathematical point of view, the coefficients of the linear combination are calculated from a system based in the following relationship established for each wavelength:

$$A^u_{\lambda_j} = \sum_{i=1}^{p} a_i A^{ref_i}_{\lambda_j} + r_j \tag{2.26}$$

where A_u and A_{ref} are, respectively, the absorbance values of the sample spectrum and of the reference spectrum, for a given wavelength λ_j, a_i is the coefficient of reference spectra contribution for the explanation of the sample spectrum and r_j is the error value.

The validation of the model is given by the sum of the error values at each wavelength, which must be as low as possible. Moreover, the variation of the error value with wavelength must also be considered (a random distribution being waited).

Starting from Eq. (2.26), any additive parameter (TOC, for example) can thus be calculated by using the following equation:

$$P_u = \sum_{i=1}^{p} a_i P_i \tag{2.27}$$

where P_u and P_i are, respectively, the parameter values of the sample and of the reference spectra, and a_i the previously calculated coefficient.

This methodology will largely be used in Chapters 7 to 10 dealing with environmental application.

References

1. S. Vaillant, M-F. Pouet, O. Thomas, *Urban Water*, 4 (2002) 273.
2. O. Thomas, *Métrologie des Eaux Résiduaires*, Cebedoc-Lavoisier/Tec et Doc (Ed), Liège-Paris (1995).
3. C. Muret, M-F. Pouet, E. Touraud, O. Thomas, *Water Sci. and Technol.*, 42 (5–6) (2000) 47.
4. J. Tölgeyssy, *Chemistry and Biology of Water Air and Soil: Environmental Aspects.* Elsevier (Ed), Amsterdam (1994).
5. O. Thomas, F. Théraulaz, C. Agnel, S. Suryani, *Envir. Technol.*, 17 (1996) 251.
6. E. Naffrechoux, N. Mazas, O. Thomas, *Environ. Technol.*, 12 (1991) 325.
7. D.V. Stynes, *Inorg. Chem.*, 14 (1975) 453.
8. E. Baures, *Non Parametric Measurement for the Study of Industrial Wastewaters.* PhD thesis, University Aix Marseille III, France (2002).
9. O. Thomas, E. Baures, M-F. Pouet, *Wat. Qual. Res. J. Canada*, 40 (1) (2005), 51.
10. S. Gallot, O. Thomas, *Fresenius J. Anal. Chem.*, 346 (1993) 976.
11. S. Vaillant, *Urban Wastewater Organic Matter: Characterization and Evolution*, PhD thesis, University of Pau, France (2000).
12. M-F. Pouet, E. Baures, S. Vaillant, O. Thomas, *Applied Spectroscopy*, 58 (4) (2004) 46.
13. S. Vaillant, M-F. Pouet, O. Thomas, *Talanta*, 50 (1999) 729.
14. J. Mendham, R.C. Denney, J.D. Barnes, M.J.K. Thomas, *Vogel's Textbook of Quantitative Chemical Analysis*, 6th edition, Prentice Hall, (2000).
15. APHA, AWWA, WEF, *Standard Methods for the Examination of Water and Wastewater*, 18th edition, A.E. Greenberg, L.S. Clesceri, A.D. Eaton (Eds), NW Washington, (1992).
16. H.H. Willard, L.L. Merritt, J.A. Dean. *Instrumental Methods of Analysis*, Van Nostrand, New York, (1974).
17. M. Blanco, J. Gene, H. Iturriaga, S. Maspoch, J. Riba, *Talanta*, 34 (1987) 987.
18. A. Cladera, E. Gómez, J.M. Estela, V. Cerda, *Anal. Chim. Acta*, 267 (1992) 95.
19. A.R. Hawthorne, S.A. Morris, R.L. Moody, R.B. Gammage, *J. Envir. Sci. Health*, A19 (1984) 253.
20. M.A. Ferree, R.D. Shannon, *Wat. Res.*, 35 (1) (2001) 327.
21. O. Thomas, S. Gallot, *Fresenius J. Anal. Chem.*, 338 (1990) 234.
22. O. Thomas, S. Gallot, N. Mazas, *Fresenius J. Anal. Chem.*, 338 (1990) 238.
23. O. Thomas, S. Gallot, E. Naffrechoux, *Fresenius J. Anal. Chem.*, 338 (1990) 241.
24. A. Oumedjbeur, O. Thomas, *Analusis*, 17 (4) (1989) 221.
25. W.H. Lawton, E.A. Sylvestre, *Technometrics*, 13 (1971) 617.
26. D.W. Osten, B.R. Kowalski, *Anal. Chem.*, 56 (1984) 991.
27. B. Vandeginste, R. Essers, T. Bosman, J. Reijen, G. Kateman, *Anal. Chem.*, 57 (1985) 971.
28. H.A. Msimanga, P.E. Sturrock, *Anal. Chem.*, 62 (1990) 2134.
29. N.D. Lourenc, C.L. Chaves, J.M. Novais, J.C. Menezes, H.M. Pinheiro, D. Diniz, *Chemosphere*, (2006) available online.

UV-Visible Spectrophotometry of Water and Wastewater
O. Thomas and C. Burgess (Eds.)
© 2007 Elsevier. All rights reserved.

CHAPTER 3

Organic Constituents

C. Gonzalez[a], E. Touraud[a], S. Spinelli[a], O. Thomas[b]

[a]Laboratoire Génie de l'Environnement Industriel, Ecole des Mines d'Alès, 6 Avenue de Clavières, 30319 Alès Cedex, France; [b]Observatoire de l'Environnement et du Développement Durable, Université de Sherbrooke, Sherbrooke, Québec, J1K 2R1, Canada

1. INTRODUCTION

The importance of identifying organic constituents in water and wastewater is related to their potential toxicity, biodegradability and availability. On the contrary, the evolution and potential toxicity of mineral constituents are generally less important with respect to their environmental impact. Although some aggregate parameters may be used for the estimation of pollution effects and treatment (see Chapter 5), the determination of specific organic compounds is necessary, namely in the frame of regulation.

Since the beginning of this century, synthetic organic compounds have been produced either for domestic (detergents, plastics, etc.), industrial (solvents, additives, dyes, etc.) or agricultural (pesticides, etc.) uses. Around 60,000 compounds are widely used in human activities and could be found in the environment, especially in water (surface water, groundwater, industrial or urban wastewater) or polluted soils [1–4].

Most of them present a conjugated structure (aromatic rings), which induces UV absorption, sometimes in the visible range [5] if the conjugation in the molecule is extended (dyes). Moreover, aliphatic compounds that have a chromophore in their structure (carbonyl group, for example) show poor absorption in the UV region. Derivatisation, with the use of a specific reagent, may enhance their absorption. Finally, some of them, such as carbohydrates, do not absorb either in visible range or in UV range (200–400 nm). However, they can be revealed after a photo-oxidation step.

2. COLOURED ORGANIC COMPOUNDS

Coloured organic compounds absorb UV-visible light, generally with a strong absorptivity in the visible range ($\varepsilon > 10^3$ Lmole^{-1}cm^{-1}). In this part are presented some dyes possessing acid–base properties, and some usual coloured reagents such as pH, redox or complexometry indicators. A careful spectroscopic study, dealing with the nature of the electronic transitions and the location of the corresponding absorption bands, is proposed.

2.1. Dyes

Dyes generally contain two or more cyclic rings that may or may not be aromatic and condensed. From a chemical point of view, a dye molecule can be characterised, on the one hand, by the basic structure, which is related to a dye family and contains chromophores (conjugated double bonds, aromatic rings), which induce the dye solution coloration, and, on the other hand, by the substituents or auxochromic groups, which infer aqueous solubility by ionisation (NH_2, OH, COOH, SO_3H, etc.) and can enhance conjugation in the dye molecule.

The most important families of dyes are azoic and anthraquinonic ones. Azoic dyes are characterised by an azo bond (N=N) connected to aromatic rings or heterocycles, meanwhile anthraquinonic dyes are derivatives of substituted anthraquinone and have two carbonyl groups (C=O) in their structure. Various substitutes can be found, such as alkyl, amino, hydroxy, halogeno, sulphonate or more complex groups. The studied dye solutions have been prepared in water at a concentration of 50 mgL^{-1}. The effect of pH on UV-visible spectra is pointed out.

2.1.1. Azoic dyes

Two isomeric phenylazonaphthols (called Orange 1 and Orange 2) and an aminobenzene derivative (Orange 3 or methyl orange) are presented.

The chemical structure of these phenylazonaphthols shows that these compounds give rise to a tautomerism between the azo and hydrazone forms by a proton exchange effect (Figs. 1 and 2). In aqueous medium, the hydrazone form is preponderant [6].

Orange 1 and Orange 2 are pH-dependent compounds. In Fig. 3, UV-visible spectra are given according to different pH values. In acid medium, the absorption band in the visible region, imputed to the $\pi-\pi^*$ transition of the hydrazone form, is observed, respectively, at 475 nm ($\varepsilon = 13950$ $Lmol^{-1}cm^{-1}$) for Orange 1 and at 483 nm ($\varepsilon = 17970$ $Lmol^{-1}cm^{-1}$) for Orange 2. The other peaks, located around 230, 240 and 270 nm, are assigned to the $\pi-\pi^*$ transitions of aromatic rings (benzene and naphthalene rings). Moreover, for Orange 2 dye, a slight shoulder is noticed around 400 nm due to the $n-\pi^*$ transition of N=N azo group. In both cases, the coloration is orange.

In basic medium, a bathochromic shift ($\lambda = 513$ nm, $\varepsilon = 14100$ $Lmol^{-1}cm^{-1}$) can be noticed for the Orange 1: the colour turns from orange to red. The phenate form is

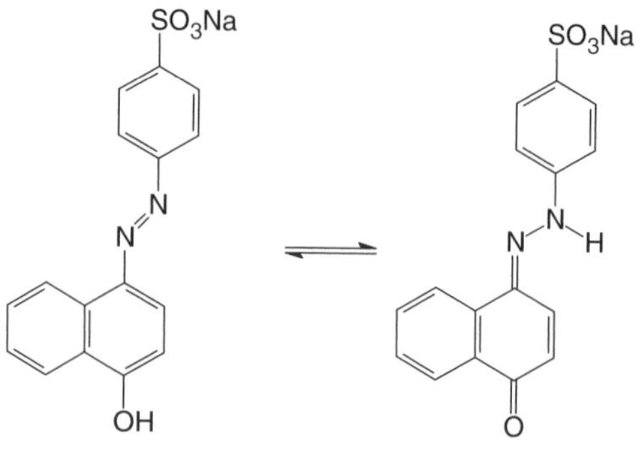

FIGURE 1. Orange 1 dye azo-hydrazone tautomerism.

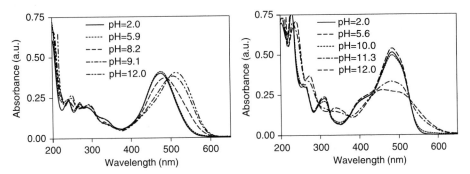

FIGURE 2. *Orange 2 dye azo-hydrazone tautomerism.*

FIGURE 3. *Orange 1 (left) and Orange 2 (right) dyes UV-visible spectra (water, pathlength: 2 mm).*

FIGURE 4. *Acid–basic equilibrium of Orange 1 dye.*

predominant (Fig. 4). The presence of an isosbestic point ($\lambda = 493$ nm) confirms this phenomenon.

In basic medium, the Orange 2 dye behaviour is a little different. The phenate ion is not easily formed because of stabilisation of the hydrazone form by a hydrogen bond between the oxygen atom of the carbonyl group and the hydrogen atom connected to the β nitrogen of the azo group (Fig. 2). Nevertheless, a change of coloration (from orange to red) is observed in strong basic medium (pH = 12.0). The shoulder around 400 nm disappears, and the absorption band in the visible region extends (aggregation process, dimerisation).

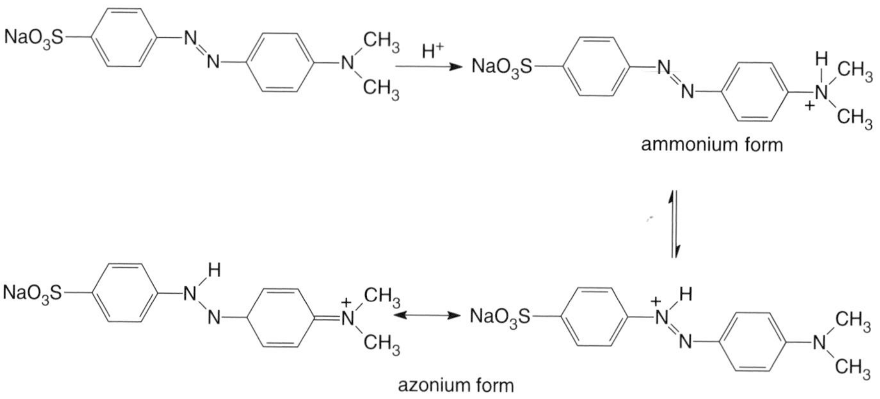

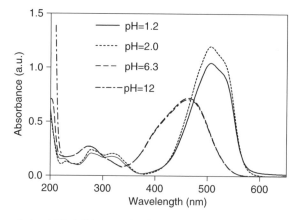

FIGURE 5. *Ammonium–azonium equilibrium of methyl orange dye.*

Orange 3 or methyl orange has not the same behaviour as the previous dyes: no azo-hydrazone tautomerism is possible. Nevertheless, in strong acid medium, an ammonium–azonium equilibrium occurs, as shown in Fig. 5.

This phenomenon has been pointed out on UV-visible spectra (Fig. 6). The maximum absorption of Orange 3 molecular form is located at 463 nm ($\varepsilon = 23{,}560$ Lmol^{-1}cm^{-1}) and imputed to the π–π* transition of azo group and the slight shoulder around 400 nm to the n–π* transition of the same group (pH = 6.3).

At pH = 2.0, the azonium tautomeric forms become preponderant and strongly absorb in the visible region ($\lambda = 506$ nm, $\varepsilon = 39{,}180$ Lmol^{-1}cm^{-1}). The colour of the solution is red.

A shoulder can be noticed also at 319 nm ($\varepsilon = 6760$ Lmol^{-1}cm^{-1}), related to the ammonium form. In a strong acid medium, it can be expected that a diprotonated form [$-^+$NH$-$(CH$_3$)$_2$, $-$N$_\alpha$ $=^+$N$_\beta$H$-$] may exist, showing an absorption at 410 nm [7].

FIGURE 6. *Orange 3 dye UV-visible spectra (water, pathlength: 2 mm).*

At pH = 1.2, an absorbance decrease at 506 nm and a very slight absorbance increase around 400 nm are observed.

As expected, no bathochromic effect is observed between neutral and basic media.

2.1.2. Anthraquinonic dyes

Four anthraquinonic dyes, Alizarin red S, Alizarin violet R, Acid green 25, and Acid blue 129 are presented. The simplest one, the Alizarin red S, is derived from alizarine by the introduction of a sulphonate group in the alizarine structure in position 3 (Fig. 7).

UV-visible spectra of Alizarin (Fig. 8) and Alizarin red S (Fig. 9) are acquired according to different pH. Alizarin is slightly soluble in water for acid pH. Sulphonic acid group induces aqueous solubility for alizarin red S.

Dealing with the solution colour, two changes can be observed for Alizarin: the colour turns progressively from pale yellow (pH = 5.2) to pale rose (pH = 7.1) and then to violet (pH = 10.1).

Alizarin exhibits two ionisable enolic functions: it can be supposed that they are ionised in strong basic medium. The bathochromic shift is then imputed to the extended conjugation of the molecule. A fine structure, imputed to the n–σ^* transition of the hydroxy groups, is noticed at pH = 12.0 (λ = 566 nm, ε = 7000 Lmol^{-1}cm^{-1} and λ = 608 nm, ε = 6500 Lmol^{-1}cm^{-1}). In the same time, the absorption peak located at 326 nm decreases.

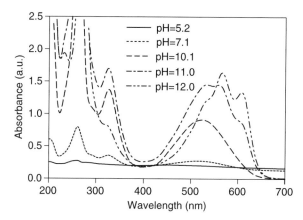

FIGURE 7. *Alizarin and Alizarin red S monohydrate chemical structures.*

FIGURE 8. *Alizarin UV-visible spectra according pH medium (water, pathlength: 10 mm).*

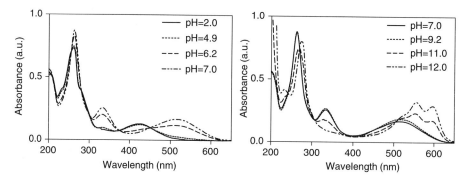

FIGURE 9. *Alizarin red S monohydrate dye UV-visible spectra (water, pathlength: 2 mm).*

At pH < 10.0, only one phenol function is ionised (probably at position 1): the two absorption peaks (λ = 326 nm and λ = 520 nm) can reasonably be assigned to the n–σ* transition of the protonated and deprotonated forms of hydroxy groups, respectively.

The absorption in UV region is imputed to the π–π* transitions of the anthraquinonic structure (benzenoid and quinonoid bands).

Similarities are observed with Alizarin red S in basic medium (Fig. 9): the same fine structure can be observed (λ = 556 nm, ε = 11600 Lmol^{-1}cm^{-1} and λ = 596 nm, ε = 10,450 Lmol^{-1}cm^{-1}). In comparison with Alizarin, a slight hypsochromic effect occurs, probably due to the sulphonic acid function. The solution is violet. Between pH = 9.0 and pH = 11.0, the colour turns from red (λ = 514 nm) to violet: an isosbestic point, located around 510 nm, shows that we are faced with two different forms (monoionised and dionised) of the molecule. A hyperchromic effect can be noticed too at pH = 12.0.

In acidic medium, two isosbestic points are pointed out at 375 and 450 nm: the colour turns progressively from pale yellow (λ = 425 nm, pH = 2.0) to yellow–orange (pH = 4.9) and then to red (λ = 510 nm, pH = 6.2). At pH = 2.0, only the molecular form is predominant (λ = 426 nm, ε = 4400 Lmol^{-1}cm^{-1}). The absorption band located at 261 nm can be assigned to the π–π* transition of quinonic structure.

With the two disubstituted anthraquinonic dyes (Green acid 25 and Violet alizarin R), the influence of the position of the substituents on the UV-visible absorption can be discussed. Green acid 25 is a 1,4 disubstituted anthraquinone, while Violet alizarin R is a 1,5 one (Fig. 10).

FIGURE 10. *Chemical structures of acid green 25 (right) and violet alizarin R (left).*

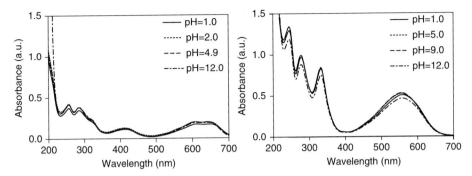

FIGURE 11. *UV-visible spectra of acid green 25 (left) and Alizarin violet R (right) dyes in acid and basic media (water, pathlength: 2 mm for acid green 25, 10 mm for Alizarin violet R).*

According to their structure, these isomeric dyes must be nonsensitive to pH. UV-visible spectra confirm this hypothesis (Fig. 11).

The absorption bands in the visible region of the spectra are attributed to resonance between the quinonic ring and the substituents. In 1,4 disubstituted anthraquinones, an interaction between the two chromophoric systems occurs, giving a resonance effect between two extreme structures. This difference in behaviour has been pointed out: 1,4 dihydroxy and diamino compounds show a "double-headed" peak that is not shown by 1,5 compounds. The latter ones behave as twice the corresponding monosubstituted compound (the molecule would than be divided diagonally) (Fig. 12) [8].

A marked bathochromic effect is observed for Acid green 25 but the usual double-headed peak is mitigated, probably because of benzoylation of amino groups.

The decreasing of the molecule's quinonoid character is associated with the visible band's weakness (Table 1).

In the UV range, benzenoid and quinonoid bands are more intense for Violet alizarin R.

Finally, the case of a 1,2,4 trisubstituted anthraquinone derivative (Acid blue 129) is presented in Fig. 13.

FIGURE 12. *1,4 and 1,5 disubstituted anthraquinones behaviour.*

C. Gonzalez et al.

TABLE 1. *Visible absorption of disubstituted anthraquinones*

Dye	λ_{max} (nm)	ε_{max} (Lmol^{-1}cm^{-1})
Acid green 25 (pH = 4.9)	610, 645	2480, 2480
Alizarin violet R (pH = 5.0)	557	6520

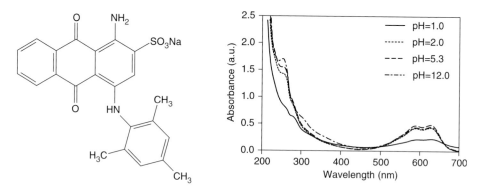

FIGURE 13. *Acid blue 129 chemical structure and UV-visible spectra (water, pathlength: 10 mm).*

The typical double-headed peak of 1,4 disubstituted derivatives is present on acid blue 125 visible range of UV spectra (λ = 588 nm and λ = 628 nm). In strong acid medium (pH = 1.0), protonation can occur on the nitrogen atom: it seems to be more easy with Acid blue 129 dye than with Acid green 25 and Alizarin violet R dyes, where the steric hindrance around nitrogen atoms is more important.

2.1.3. Other dyes

There exist other dyes, among which are derivatives of triphenylmethane. These dyes usually possess acid–base indicator properties. Phenol red belongs to this family. Its structure gives rise to different equilibria according to pH (Fig. 14). In strong acid medium (pH = 1.0), there is protonation of tautomeric forms Ia and Ib to give form II. In more basic medium, the equilibrium phenol (form I)/phenate (form III) is observed (pK$_a$ = 8.0).

UV spectrophotometry allows pointing out these different forms (Fig. 15). In acid medium, forms I and II coexist; form II prevails at pH = 1.0 and presents a strong absorption in the visible region (λ = 504 nm, ε = 42,300 Lmol^{-1}cm^{-1}). At pH = 3.6, the colour of the solution turns from orange to yellow: the main absorption of form Ia and Ib is located at 432 nm (ε = 20,640 Lmol^{-1}cm^{-1}). An isosbestic point can be noticed at 466 nm.

In basic medium, the phenate form is predominant (pK$_a$ = 8.0); both bathochromic and hyperchromic effects are observed for the transition n–σ* of hydroxy group (λ = 558 nm, ε = 56,540 Lmol^{-1}cm^{-1}). The coloration turns red. Another specific isosbestic point appears at 481 nm.

In the UV range, absorption bands around 270 nm can be reasonably imputed to π–π* transition of aromatic rings.

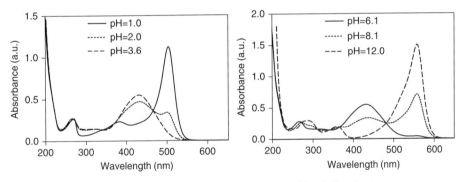

FIGURE 14. *Phenol red equilibria according to pH medium.*

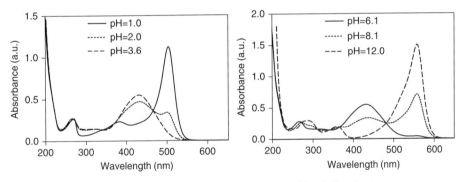

FIGURE 15. *Phenol red UV-visible spectra according to pH solution (water, pathlength: 2 mm).*

Crystal violet is another triphenylmethane derivative for which a tautomeric and pH-dependent equilibrium exists between the triphenylmethyl cation and its quinoidal form (Fig. 16). In acid medium, the quinoidal form leads to mono and dicationic forms by protonation on nitrogen atoms (Fig. 17). The pH range for Crystal violet is 0.0–1.8. UV spectrophotometry points out the three different forms (Fig. 18).

At pH = 5.3, the neutral form is predominant and absorbs strongly in visible range (λ = 590 nm, ε = 81000 Lmol^{-1}cm^{-1}). The solution is dark violet. No change occurs in basic medium but, at pH = 12.0, the solution loses colour in few minutes. In strong acid medium acid (pH = 1.0), monocationic form (λ = 425 nm, ε = 13500 Lmol^{-1}cm^{-1})

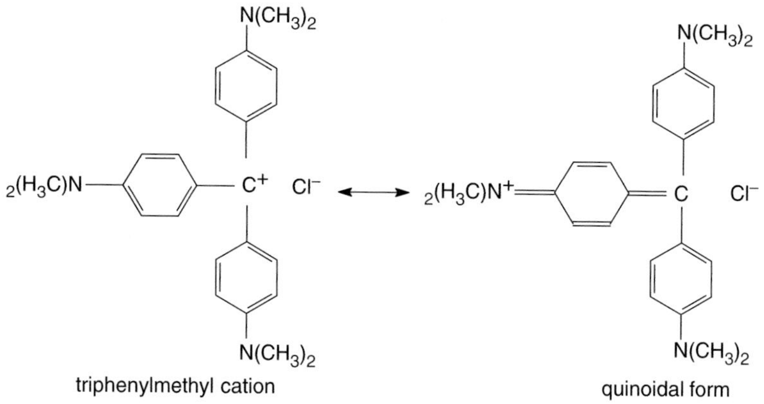

FIGURE 16. *Crystal violet tautomerism.*

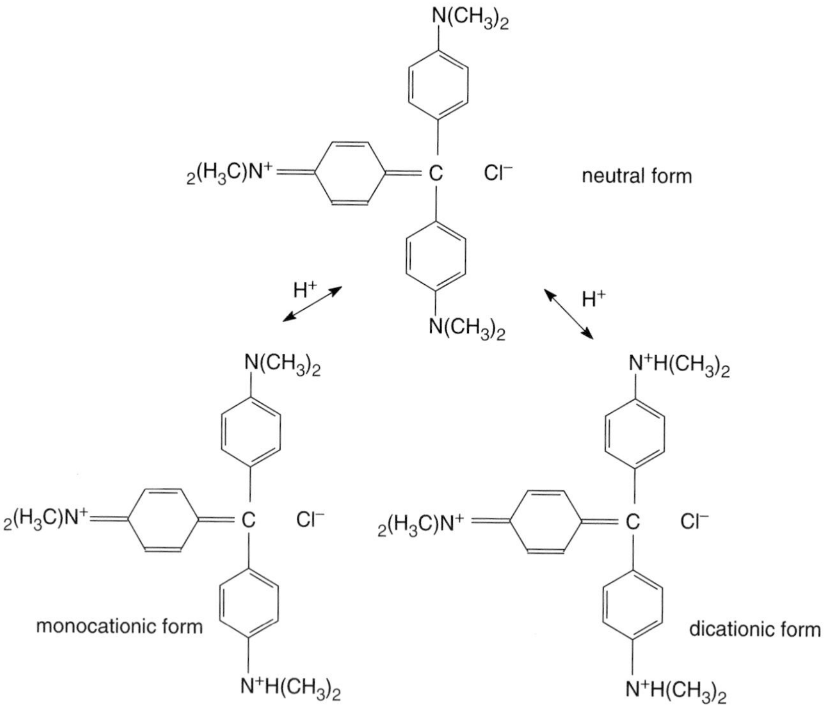

FIGURE 17. *Crystal violet mono and dicationic forms in acid medium.*

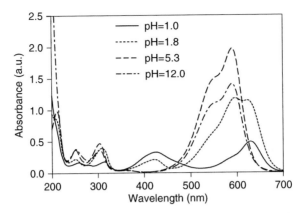

FIGURE 18. *Crystal violet UV-visible spectra according to pH medium (water, pathlength: 2 mm).*

and dicationic form ($\lambda = 630$ nm, $\varepsilon = 20100$ Lmol^{-1}cm^{-1}) co-exist; the colour of the solution is green, between yellow and blue. At pH $= 1.8$, a mixture of the neutral and monocationic forms is present.

2.2. Coloured reagents

The studied reagents solutions have been prepared in water at a concentration of 50 mgL^{-1}.

2.2.1. pH indicators

Three pH indicators, the pH range of which varies from 4 to 12, are presented: Methyl red, Alizarin yellow R, and Bromothymol blue.

Methyl red (pH range: 4.4–6.2) is an azo compound, the structure of which differs from methyl orange (Orange 3) one by the substitution of sulphonic acid function by carboxylic acid function (Fig. 19).

UV-visible spectra are displayed according to pH (Fig. 20). As for Orange 3, it can be supposed that, in strong acid medium, an ammonium–azonium equilibrium occurs.

At pH $= 1.0$, the azonium forms are predominant and strongly absorb in the visible range ($\lambda = 515$ nm, $\varepsilon = 23,000$ Lmol^{-1}cm^{-1}). Between pH $= 2.0$ and 7.0, the aqueous solubility of the compound is incomplete, indicating a partial ionisation of the carboxylic

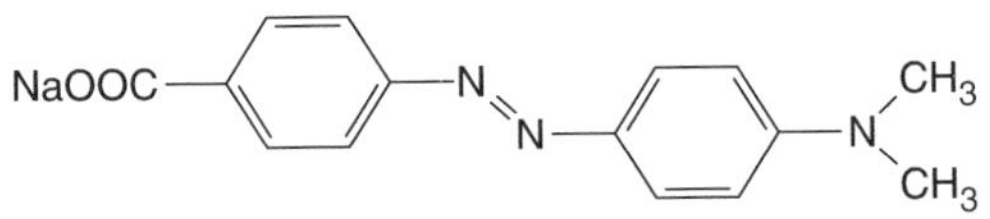

FIGURE 19. *Methyl red chemical structure.*

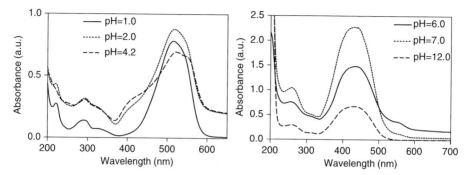

FIGURE 20. *Methyl red UV-visible spectra according to pH medium (water, pathlength: 2 mm at pH = 1.0 and 12.0, 10 mm at pH = 2.0, 4.2, 6.0 and 7.0).*

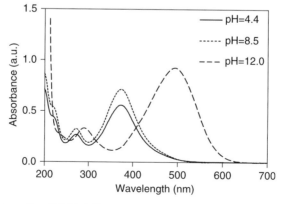

FIGURE 21. *Alizarin yellow R chemical structure.*

function (pK$_a$ = 5.0). At pH > 7.0, the maximum of absorption is located at 431 nm (ε = 19600 Lmol^{-1}cm^{-1} at pH = 12.0): the coloration of the solution is yellow.

The chemical structure of Alizarin yellow R (pH range = 10.1–12.0) is given in Fig. 21. This azo compound is very slightly soluble in water in strong acid medium (pH = 1.0 or 2.0) where carboxylic acid and phenol functions are not ionised. At pH = 4.4, the aqueous solubility increases due to the ionisation of carboxylic group. In basic medium, the phenate form is predominant.

UV-visible spectra of Alizarin yellow R, according to pH (after filtration at pH = 4.4), are shown in Fig. 22. A strong bathochromic effect can be noticed at pH = 12.0. At pH = 8.5,

FIGURE 22. *Alizarin yellow R UV-visible spectra according to pH medium (water, pathlength: 10 mm at pH = 4.4, 2 mm at pH = 8.5 and 12.0).*

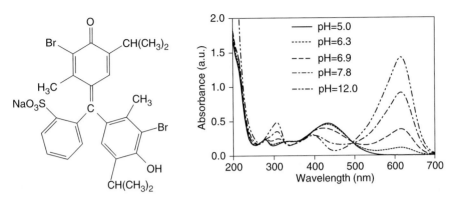

FIGURE 23. *Bromothymol blue chemical structure and UV-visible spectra (water, pathlength: 10 mm).*

the colour of the solution turns from yellow ($\lambda = 373$ nm, $\varepsilon = 20500$ Lmol^{-1}cm^{-1}) to red at pH $= 12.0$ ($\lambda = 493$ nm, $\varepsilon = 27000$ Lmol^{-1}cm^{-1}).

Bromothymol blue (pH range: 6.2–7.6) is a triphenylmethane derivative, the chemical structure and spectral behaviour of which are close to the ones of phenol red (Fig. 23). In basic medium, a strong bathochromic and hyperchromic effect can be noted ($\lambda = 615$ nm, $\varepsilon = 17800$ Lmol^{-1}cm^{-1}). The colour turns from yellow ($\lambda = 433$ nm, $\varepsilon = 5900$ Lmol^{-1}cm^{-1}, pH $= 5.0$) to blue (pH $= 12.0$). The bromine atom located on the phenolic ring induces a more intense bathochromic shift than for Phenol red.

2.2.2. Redox indicator

Ferroin or tri (1,10-phenanthrolin) iron II is not pH dependent, as shown by UV visible spectra (Fig. 24).

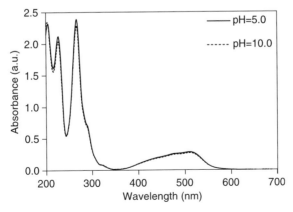

FIGURE 24. *Ferroin UV-visible UV spectra in acid and basic media (water, pathlength = 10 mm).*

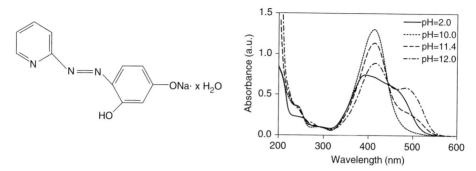

FIGURE 25. *PAR monosodium salt hydrate UV-Visible spectra according to pH (water, pathlength: 2 mm).*

The extended conjugation in the molecule due to the phenanthrolin structure induces an absorption in the visible range: the solution coloration is orange.

2.2.3. Complexometry indicators

Three indicators used in complexometric determinations of metals are presented (4-(2-pyridylazo) resorcinol (PAR), Eriochrom black T and Dithizone or diphenylthiocarbazone).

The 4-(2-pyridylazo) resorcinol, monosodium salt hydrate (PAR, monosodium salt) is a reagent for selective determination of metallic compounds such as Cr (III). Its chemical structure and UV-visible absorption are given in Fig. 25. As for phenylazonaphthols, PAR molecule gives rise to a tautomerism between the azo and hydrazone forms, the latter being predominant in water. In acid medium (pH = 2.0), a large absorption band occurs in visible range (yellow–orange). In neutral and basic media, the maximum of absorption is located at 413 nm ($\varepsilon = 31000$ Lmol^{-1}cm^{-1}, pH = 10.0). At pH = 12.0, the coloration turns from yellow to orange ($\lambda = 482$ nm, $\varepsilon = 14000$ Lmol^{-1}cm^{-1}); the two phenol functions of resorcinol are ionised.

The Eriochrom Black T is an azo compound, the tautomerism equilibrium of which is presented in Fig. 26.

UV-visible spectra according to pH are shown in Fig. 27. In the visible range, the absorption bands are rather large with slight shoulders. A maximum of bathochromic shift can be noticed at pH = 9.0–10.0.

Dithizone is also used for the determination of metallic compounds such as copper (II) (Fig. 28). Because of the slight aqueous solubility of this compound, UV-visible spectra have been acquired in methanol/water (50/50) solution (Fig. 29). In strong acid medium (pH = 2.0), the maximum of absorption in the visible region is located at $\lambda = 588$ nm ($\varepsilon = 13,700$ Lmol^{-1}cm^{-1}). A protonation may occur on the sulphur atom. The coloration of the solution is dark blue. In basic medium (pH = 12.0), the colour turns to orange ($\lambda = 472$ nm, $\varepsilon = 8100$ Lmol^{-1}cm^{-1}). Two isosbestic points, located at $\lambda = 425$ nm and $\lambda = 517$ nm, confirms this phenomenon.

3.   UV-ABSORBING ORGANIC COMPOUNDS

Some of UV-absorbing organic compounds are known for their toxic or organoleptic effects. According to their toxicity, international legislation has established several lists

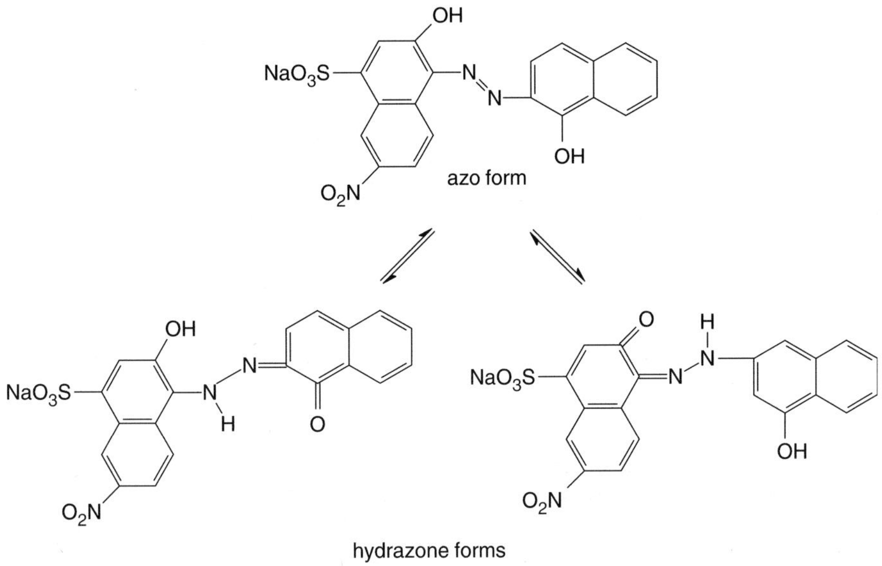

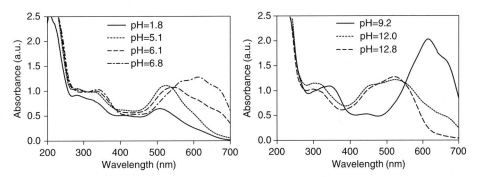

FIGURE 26. *Eriochrome Black T azo-hydrazone tautomerism.*

FIGURE 27. *Eriochrome Black T UV-visible spectra according to pH (water, pathlength: 10 mm).*

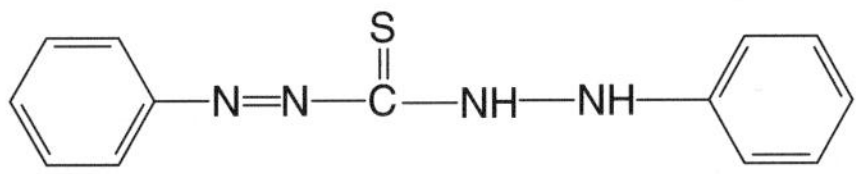

FIGURE 28. *Dithizone chemical structure.*

for the limitation of their use. These checklists include pesticides, chlorinated solvents, polycyclic aromatic hydrocarbons and phenols [9].

Moreover, a lot of organic compounds are very often used as solvents or reagents in industrial processes. These compounds (acids, aldehydes, ketones, etc.) can however be at the origin of accidental pollution if their concentration is high in industrial wastewater

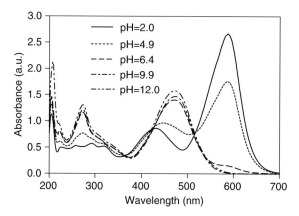

FIGURE 29. *Dithizone UV-visible spectra according to pH (methanol/water: 50/50, pathlength: 10 mm).*

discharges [10]. According to those reasons, the organic compounds considered in this section are the following:

- aldehydes and ketones
- amines
- benzene and related compounds (BTEX)
- polycyclic aromatic hydrocarbons (PAH)
- pesticides
- phenols
- phtalates
- sulphur organic compounds
- surfactants

All these compounds are considered in the final library of UV spectra (Chapter 11).

3.1. Aldehydes and ketones

In general, carbonyl compounds (aldehydes, ketones) present a poor absorption in the UV region, except benzaldehyde (Table 2).

3.1.1. Aldehydes

As previously noted, formaldehyde does not present a significant absorption spectrum (without derivatives, see Section 4.1). On the contrary, acetaldehyde, butyraldehyde and benzaldehyde show absorption maxima of different intensities according to absorptivities (Fig. 30). For benzaldehyde, the peak position is close to the one of aromatic rings.

3.1.2. Ketones

Like aldehydes, ketones generally present an absorption in the UV region due to the carbonyl group. For example, acetone and butanone have an absorption maximum at

TABLE 2. *Absorptivities of some carbonyl compounds*

Compounds	Wavelength (nm)	ε (Lmol^{-1}cm^{-1})
Formaldehyde	200	0.2
Acetaldehyde	277	6.0
Butyraldehyde	283	47
Benzaldehyde	251	12571
Acetone	266	20
2-butanone	268	17
Methyl isobutyl ketone	248	27

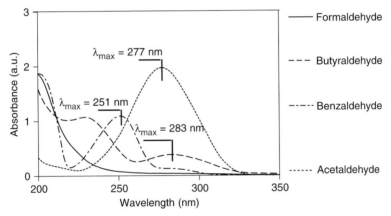

FIGURE 30. *UV spectra of formaldehyde (300 gL^{-1}), acetaldehyde (12 gL^{-1}), butyraldehyde (10 gL^{-1}) and benzaldehyde (10 mgL^{-1}).*

266 nm and 268 nm, respectively (Fig. 31). The difference between the absorption peak position of diisobutylketone (ramified ketone) and butanone is approximately 20 nm.

3.2. Amines

Aromatic amines (aniline and substituted anilines) are used as intermediates in industrial and pharmaceutical chemistry [11]. Because of their aromatic character, they strongly absorb in the UV range. A recent study has shown the great interest for the study of aromatic amines from azo-dyes reduction using UV spectrophotometry [12]. On the opposite, aliphatic amines do not absorb directly (see Section 4.2).

3.2.1. Aniline

Aniline dissociation equilibrium shows that the dissociated form prevails for pH lower than the pKa value and the undissociated form for pH greater than the pKa value (Fig. 32).

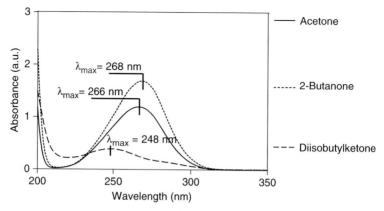

FIGURE 31. *UV spectra of acetone (4 gL^{-1}), 2-butanone (6 gL^{-1}) and diisobutylketone (500 mgL^{-1}).*

FIGURE 32. *Dissociation equilibrium of aniline.*

The UV spectrum of a basic aniline solution presents two maxima (230 nm and 280 nm), whereas in acidic conditions, the spectrum shape does not show any specific absorption band and is not exploitable. For a pH of 4.3, close to the pKa value (4.6), both dissociated and undissociated forms are present (Fig. 33).

3.2.2. Chloroanilines

Compared to chlorophenol (see Section 3.6), the chlorine position on the aromatic ring of chloroaniline seems to have a small influence on the position of the absorption maximum (bathochromic shift of approx. 10 nm, see Fig. 34).

Concerning disubstituted chloroaniline, the position of the absorption maximum is of the same order as the one of monochloroaniline (Fig. 35).

3.2.3. Toluidine and anisidine

No significant bathochromic shift is observed between spectra of aniline and toluidine (methyl-substituted aniline) (Fig. 36). When aniline is substituted by a methoxy group (anisidine), a bathochromic shift of 17 nm is observed. Thus, methoxy group has a greater influence than methyl group on the position band of substituted aniline.

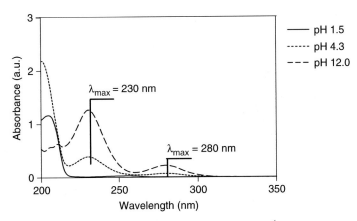

FIGURE 33. *pH effect on the UV spectrum of aniline (15 mgL⁻¹).*

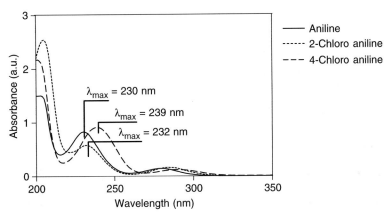

FIGURE 34. *UV spectra of monochlorinated aniline (10 mgL⁻¹).*

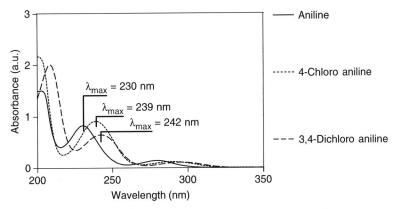

FIGURE 35. *UV spectra of aniline and dichlorinated aniline (10 mgL⁻¹).*

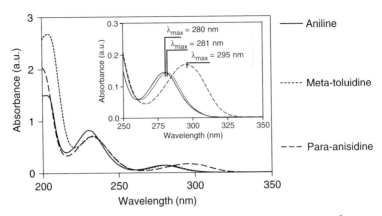

FIGURE 36. *UV spectra of aniline, toluidine and para-anisidine (10 mgL^{-1}).*

3.3. Benzene and related compounds

3.3.1. BTEX

The term BTEX concerns benzene and its alkyl derivatives such as toluene, ethylbenzene and xylenes. These compounds are semivolatile and, therefore, the solution must be prepared with great attention in order to minimise evaporation loss.

The shape of benzene and all alkylbenzenes UV spectra is characterised by a relatively fine structure. A general bathochromic effect can be observed according to the nature and position of the substituents with regard to benzene UV spectrum. The latter has a maximum at 256 nm, whereas the maximum is around 262 nm for toluene and ethylbenzene (Fig. 37) and around 266 nm for xylenes (Fig. 38). Concerning xylenes, the bathochromic shift is more important for the parasubstituted isomer (respectively, 4 and 3 nm compared to the spectra of 2-xylene and 3-xylene). In this case, a hypochromic effect is added to the bathochromic shift.

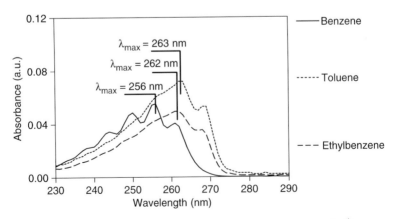

FIGURE 37. *UV spectra of benzene, toluene and ethylbenzene (40 mgL^{-1}).*

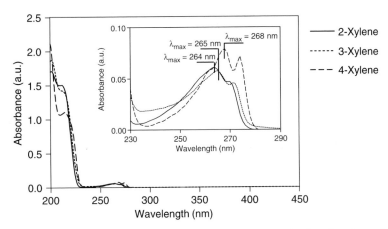

FIGURE 38. *Substitution effect of a methyl group on the aromatic ring: 2-xylene, 3-xylene, 4-xylene (30 mgL^{-1}).*

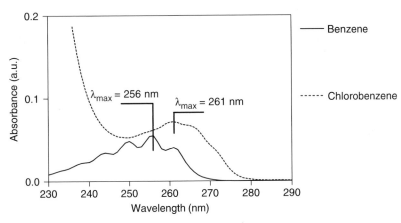

FIGURE 39. *Benzene and chlorobenzene (40 mgL^{-1}).*

3.3.2. Chlorobenzene

Benzene could also be substituted by a chlorine atom (chlorobenzene) or a nitro group (nitrobenzene). The effect of the addition of one chlorine atom on the aromatic ring leads to a bathochromic shift of 5 nm (Fig. 39).

3.4. PAH

Polycyclic aromatic hydrocarbons (PAHs) are highly toxic pollutants ($LC_{50} <$ mgL^{-1} for aquatic organisms), and some of them have proven to be carcinogenic [8,13]. They are thus monitored in potable water and wastewater. PAHs are also often present at high

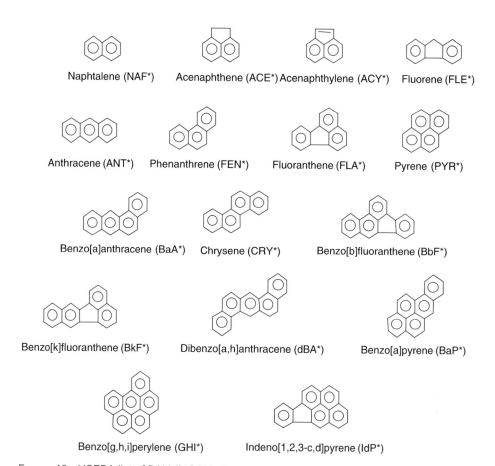

FIGURE 40. USEPA list of PAH (*: PAH abbreviated name).

concentrations in soils, i.e. on sites previously occupied by gas works and cooking plants. In both cases, as they are relatively insoluble in water, a liquid/solid extraction step is necessary, and UV spectra are performed in organic medium.

PAH are constituted by two or more aromatic rings joined together or separated by a five-membered cycle. The studied PAHs and their structures are shown in Fig. 40. In addition to these 16 PAHs from the USEPA list, benzo[b]fluorene and benzo[e]pyrene are also studied.

3.4.1. Solvent effect

Only light PAHs (2 or 3 rings) are soluble in water at the mgL^{-1} level. Figure 41 shows UV spectra of acenaphthylene in water and in acetonitrile. The shape of the two spectra is similar, and no solvent effect is observed on the absorption wavelengths. Consequently, all PAH UV spectra have been acquired in acetonitrile solutions.

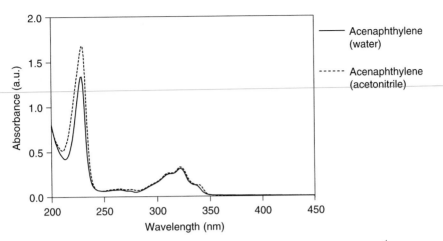

FIGURE 41. *UV spectra of acenaphthylene in water and acetonitrile (5 mgL^{-1}).*

3.4.2. Influence of the number of aromatic rings

The shape of UV spectra of PAHs is related to the number of aromatic rings and their arrangement (linear, angular or clusters) in the PAH molecule. A general bathochromic effect is observed as the number of aromatic rings increases in the PAH molecule.

For example, Fig. 42 shows UV spectra of naphthalene (two aromatic rings), phenanthrene (three aromatic rings), pyrene (four aromatic rings), benzo[a]pyrene (five aromatic rings) and benzo[g,h,i]perylene (six aromatic rings). However, the bathochromic effect weakens as the ring number increases, especially for high ring numbers as shown by benzo[a]pyrene and benzo[g,h,i]perylene.

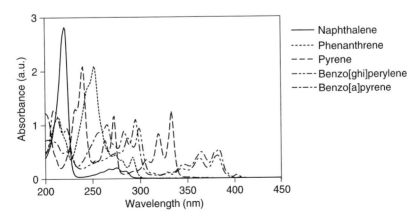

FIGURE 42. *UV spectra of naphthalene, phenanthrene, pyrene, benzo[a]pyrene and benzo[g,h,i]perylene (5 mgL^{-1}).*

C. Gonzalez et al.

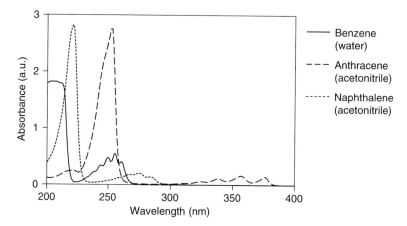

FIGURE 43. *UV spectra of naphthalene and anthracene (5 mgL^{-1}) with reference to benzene (400 mgL^{-1}).*

Dealing with the arrangement of aromatic rings in the PAH molecule, this bathochromic effect is especially important for PAHs with a linear arrangement (Fig. 43).

This phenomenon can be explained by an important delocalisation of π electrons promoted by the linear annelation. For an angular arrangement, the observed bathochromic shift is weaker (Fig. 44).

The main absorption peak wavelengths and the corresponding absorptivities for the studied PAHs are given in Table 3.

3.4.3. Isomeric PAH UV spectra

Pure aromatic PAH isomers have different UV spectra as shown in Fig. 45 (anthracene and phenanthrene) and in Fig. 46 (benzo[a]pyrene and benzo[e]pyrene). A hypsochromic

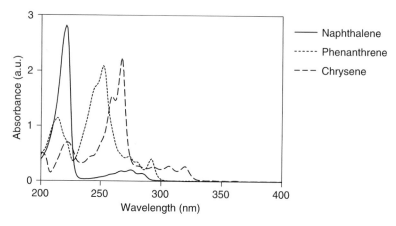

FIGURE 44. *UV spectra of phenanthrene and chrysene in reference to naphthalene (5 mgL^{-1}).*

TABLE 3. *PAH main absorption peak characteristics*

PAH name	Ring number (aromatic + nonaromatic)	Wavelength (nm)	ε (Lmol^{-1}cm^{-1})
Acenaphthene	3 (2 + 1)	289	7500
Acenaphthylene	3 (2 + 1)	322	10,000
Anthracene	3 (3)	357	6400
Benzo[a]anthracene	4 (4)	286	83,300
Benzo[a]pyrene	5 (5)	295	53,500
Benzo[b]fluoranthene	5 (4 + 1)	300	37,100
Benzo[g,h,i]perylene	6 (6)	299	53,400
Benzo[k]fluoranthene	5 (4 + 1)	307	55,800
Chrysene	4 (4)	268	106,000
Dibenzo[a,h]anthracene	5 (5)	294	110,200
Fluoranthene	4 (3 + 1)	286	27,600
Fluorene	3 (2 + 1)	263	22,700
Indeno[1,2,3-c,d]pyrene	6 (5 + 1)	302	31,200
Naphthalene	2 (2)	275	5700
Phenanthrene	3 (3)	252	63,000
Pyrene	4 (4)	333	48,300

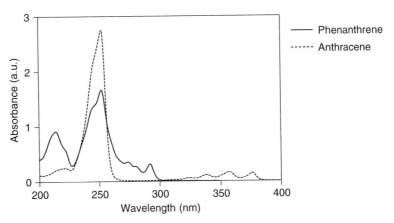

FIGURE 45. *UV spectra of anthracene and phenantrene (4 mgL^{-1}).*

shift and a hypochromic effect are observed for phenanthrene (angular arrangement) in comparison with anthracene (linear arrangement).

For benzene clusters such as benzo[a]pyrene and benzo[e]pyrene, UV spectra also reveal a hypsochromic shift and a hypochromic effect for benzo[e]pyrene which is more condensed.

The difference is not so marked for benzo[b]fluoranthene and benzo[k]fluoranthene, which have a five-membered cycle in their structure. Nevertheless, a slight bathochromic effect can be noticed for the less-condensed isomer (Fig. 47).

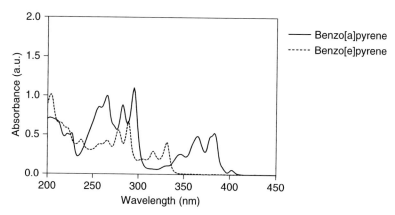

FIGURE 46. *UV spectra of benzo[a]pyrene and benzo[e]pyrene (5 mgL^{-1}).*

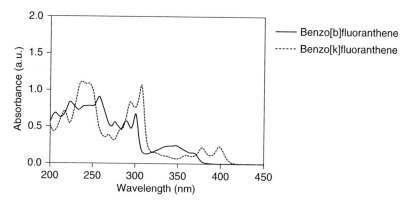

FIGURE 47. *UV spectra of benzo[b]fluoranthene and benzo[k]fluoranthene (5 mgL^{-1}).*

3.4.4. Introduction of a five-membered cycle in the PAH structure

Some PAH molecules include in their structure a five-membered cycle. When an aromatic ring is added to the linear chain containing the five-membered cycle, a bathochromic shift and hyperchromic effect are observed.

The following figures illustrate this phenomenon for fluorene and benzo[b]fluorene (Fig. 48) and fluoranthene and benzo[k]fluoranthene (Fig. 49), respectively.

When the aromatic ring is not added to the linear chain, these effects are less significant (benzo[b] fluoranthene and indeno[1,2,3,cd]pyrene). Figure 50 presents some UV spectra of nonaromatic PAHs, with an increasing ring number (from 3 to 6). In comparison with aromatic PAHs, a bathochromic effect can be noticed as the ring number increases but a general hypochromic effect is observed.

Figure 51 gives the absorption range of the 16 USEPA PAHs, showing the main and secondary peaks for each of them.

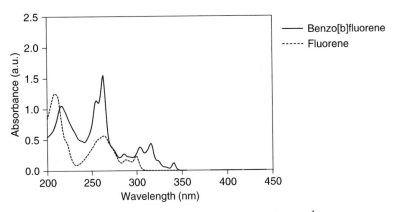

FIGURE 48. *UV spectra of fluorene and benzo[b]fluorene (5 mgL^{-1}).*

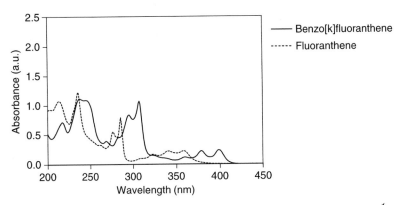

FIGURE 49. *UV spectra of and fluoranthene and benzo[k]fluoranthene (5 mgL^{-1}).*

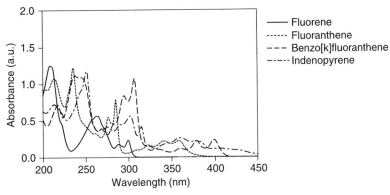

FIGURE 50. *UV spectra of nonaromatic PAHs (5 mgL^{-1}).*

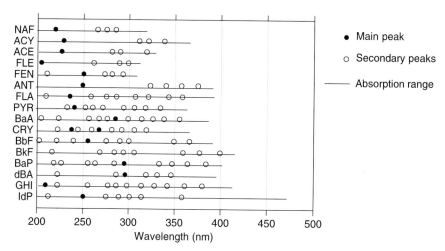

FIGURE 51. *Absorption range of the 16 USEPA PAHs.*

In conclusion, according to the aromatic character of PAH molecules, some general tendencies can be pointed out from the study of UV spectra:

- a bathochromic shift is observed as the length of the aromatic chain increases in the PAH molecule
- a hyperchromic effect is noted when aromatic rings are less condensed
- the presence of a five-membered cycle in the PAH molecule induces a bathochromic effect

3.5. Pesticides

Pesticides are usually classified according to either their chemical type or their use [14–16]:

- insecticides (organochlorinated, organophosphorous)
- herbicides (nitrogenous or nonnitrogenous)

Like PAHs, these compounds are slightly soluble in water. Therefore, solvent solutions (methylene chloride for example) must be prepared for their spectroscopic study.

3.5.1. Insecticides

Organochlorinated compounds consist of two different major groups, the cyclodiene group (heptachlor) and the DDT group (containing two aromatic rings. Many organophosphorous pesticides have been produced in order to replace more persistent organochlorinated pesticides (as DDT). However, some of these compounds are used as fungicides or herbicides.

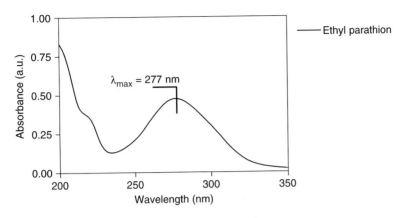

FIGURE 52. *UV spectrum of ethyl parathion (24 mgL⁻¹).*

Ethyl parathion (O,O-diethyl O-4-nitrophenyl phosphorothioate) contains a nitrophenyl group and thus presents an absorption peak around 280 nm (Fig. 52).

3.5.2. Herbicides

Phenoxycarboxylic acids (phenoxyacetic acids such as 2,4-D, Fig. 53) are the most important herbicides in terms of production.

Some nitrogenous herbicides are derived from urea, such as diuron, and contain an aromatic ring (phenylurea). The corresponding UV spectrum shows an absorption peak at 250 nm (Fig. 54).

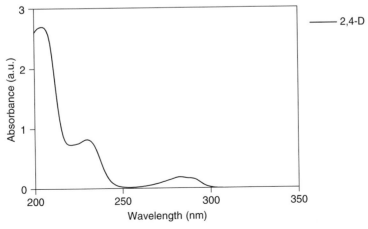

FIGURE 53. *UV spectrum of 2,4-D (20 mgL⁻¹).*

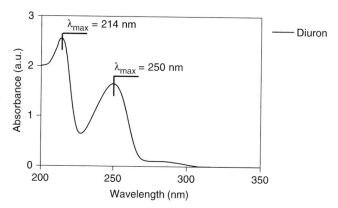

FIGURE 54. *UV spectrum of diuron (20 mgL^{-1}).*

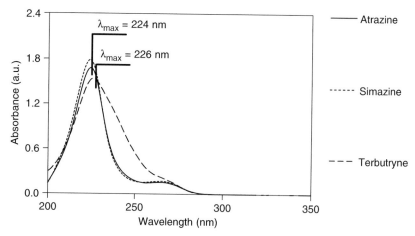

FIGURE 55. *UV spectra of atrazine, simazine and terbutryne (10 mgL^{-1}).*

The triazine herbicides are based on an *s*-triazine structure containing two amino groups. Atrazine, simazine and terbutryne present an absorption peak around 225 nm (Fig. 55), whereas hexazinone (Fig. 56) has an absorption maximum at 248 nm.

Other nitrogenous herbicides are derived from phenols and show the same UV spectrum shape as nitrophenol, such as dinoterb (terbutyl dinitrophenol, Fig. 57), for instance.

Dipyridyls are also nitrogenous herbicides. For example, paraquat is used as dichloride salt. Therefore, this compound is very soluble in water and poorly soluble in solvents. Figure 58 shows the UV spectrum of a paraquat solution.

3.6. Phenols

Several industries produce or use phenolic compounds such as alkylphenols, chlorophenols, nitrophenols, aminophenols, polyphenols or polyaromatic phenols. Chlorophenols

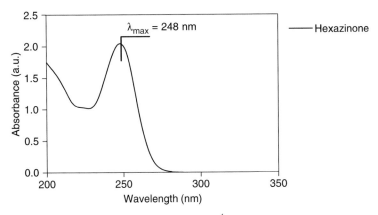

FIGURE 56. *UV spectrum of hexazinone (30 mgL^{-1}).*

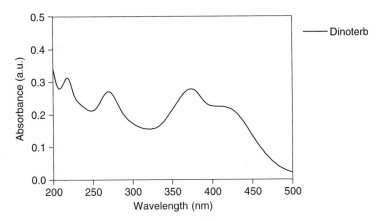

FIGURE 57. *UV spectrum of dinoterb (50 mgL^{-1}).*

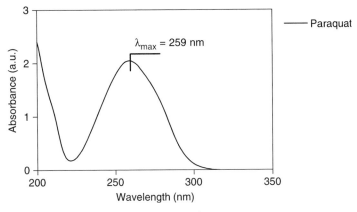

FIGURE 58. *UV spectrum of paraquat (30 mgL^{-1}).*

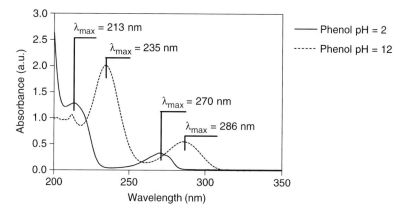

<small>FIGURE</small> 59. *Phenol dissociation equilibrium.*

<small>FIGURE</small> 60. *UV spectra of phenol (30 mgL^{-1}) in acidic and basic media.*

and nitrophenols are famous for their toxicity and also for their organoleptic properties. Owing to their acidic character, they could exit as undissociated or dissociated forms (Fig. 59):

The undissociated form is observed for pH values lower than the pK_a, and the ionised form is predominant for pH values higher than the pK_a. In these conditions, a bathochromic shift generally appears between acidic and basic media and can be used for phenolic compounds detection (Fig. 60).

The bathochromic shift, according to acidic and basic conditions, is observed for alkylphenols and chlorophenols around 19 nm (Table 4). For nitrophenols, the shift depends on the substituents number and position.

On the other hand, UV spectra (shape, bands position, absorptivities) depend on the phenolic family and substitution (number and position).

3.6.1. Alkylphenols

In this case, phenol is substituted, for example, either by a methyl group (cresols, dimethylphenol, trimethylphenol), or by a tributyl group (tributyl-4-methylphenol). The effect of an alkyl substituent induces a bathochromic shift of the characteristic phenolic band ($\lambda_{max} = 270$ m, see Fig. 60), for example, to 277 nm for 4-cresol. The para position induces the more important shift between the substituted compound and unsubstituted phenol. The previous shift is nevertheless independent of the number of methyl groups (dimethylphenol, trimethylphenol) and of the nature of the predominant forms (thus, of

TABLE 4. *Bathochromic shift between undissociated and dissociated forms*

Compound	Undissociated forms λ_{max} (nm)	Dissociated forms λ_{max} (nm)	Bathochromic shift $\Delta(\lambda_{max})$ nm
phenol	270	286	16
2-cresol	271	287	16
3-cresol	272	288	16
4-cresol	277	294	17
2,5-dimethylphenol	274	290	16
2,4,6-trimethylphenol	277	295	18
2-terbutyl 4-methyl-phenol	278	297	19
2-chlorophenol	274	292	18
3-chlorophenol	274	291	17
4-chlorophenol	279	298	19
2,3-dichlorophenol	277	297	20
2,4-dichlorophenol	283	306	23
2,4,6-trichlorophenol	286	312	26
pentachlorophenol	302	320	18
4-chloro 3-methyl phenol	279	297	18

TABLE 5. *Position of the phenol band shift calculated from phenol band, for undissociated and dissociated forms*

Compound	Undissociated form λ_{max} (nm)	Shift[1] (nm)	Dissociated form λ_{max} (nm)	Shift[2] (nm)
2-cresol	271	1	287	1
3-cresol	272	2	288	2
4-cresol	277	7	294	8
2,5-dimethylphenol	274	4	290	4
2,4,6-trimethylphenol	277	7	295	9
2-terbutyl-4-methylphenol	278	8	297	11
2-chlorophenol	274	4	292	6
3-chlorophenol	274	4	291	5
4-chlorophenol	279	9	298	12
2,3-dichlorophenol	277	7	297	11
2,4-dichlorophenol	283	13	304	20
2,4,6-trichlorophenol	286	16	312	26
Pentachlorophenol	302	32	320	34
4-chloro 3-methylphenol	279	9	297	11

(1) Shift calculated from phenol band obtained for undissociated form (270 nm).
(2) Shift calculated from phenol band obtained for dissociated form (286 nm).

the pH value). Considering the close values of the shifts (with regard to phenol) of tributyl 4-methylphenol and of 4-cresol (4 methylphenol), it is obvious that the main shift factor in this case is the para position of the methyl substituent.

Table 5 displays the values of the different shifts corresponding to the studied alkylphenols.

C. Gonzalez et al.

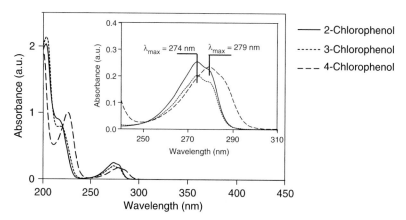

FIGURE 61. *UV spectra of monochlorophenol isomers (15 mgL⁻¹): 2-chlorophenol, 3-chlorophenol, 4-chlorophenol.*

3.6.2. Chlorophenols

For monochlorophenols, the parasubstituted isomer (4-chlorophenol) presents an absorption maximum at a wavelength higher than those of the other isomers (Fig. 61). Moreover, contrary to alkylphenols, the number of substituents has an influence on the position of the bands, increasing the shift with phenol (Table 5). Thus, pentachlorophenol leads to the greater shift value, whatever its form (undissociated or dissociated). For 4-chloro 3-methyl-phenol, the substituted methyl group seems to have no influence on the position of the band compared to the one of 4-chlorophenol.

3.6.3. Nitrophenols

As with alkyl and chlorophenols, the position of the band of para-substituted nitrophenol isomers is different from the other isomers (Fig. 62).

3.6.4. Polyphenols (catechol)

The presence of a second hydroxy group on the aromatic ring does not modify the position of the absorption maximum with regard to phenol, if the two hydroxy groups are contiguous (Fig. 63).

   Like chlorophenols, the presence of chlorine atoms on the aromatic ring induces a bathochromic shift.

3.7. Phthalates

Phthalates are often used as plasticisers for polyvinyl, polyvinyl chloride or cellulose resins. They are esters of phthalic acid. The absorption maxima of diethylphthalate, dibutylphthalate and butylbenzylphthalate are around 275 nm (Fig. 64).

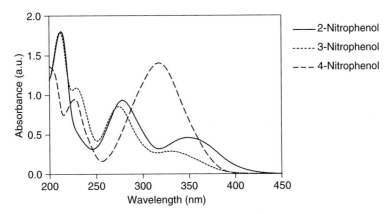

FIGURE 62. *UV spectra of nitrophenol isomers (20 mgL^{-1}), 2-nitrophenol, 3-nitrophenol, 4-nitrophenol.*

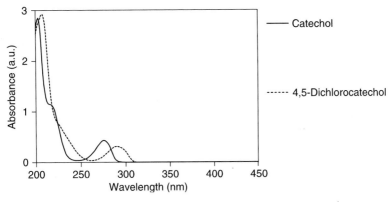

FIGURE 63. *UV spectra of catechol and 4,5 dichlorocatechol (20 mgL^{-1}).*

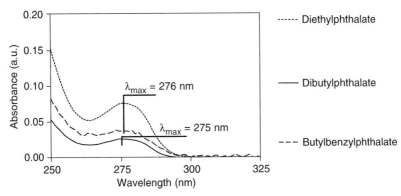

FIGURE 64. *UV spectra of phtalates (10 mgL^{-1}).*

C. Gonzalez et al.

3.8. Sulphur organic compounds

Organic sulphur compounds are volatile; a few are soluble in water and are associated with bad odour. They are essentially studied in gaseous phase. But they can be present in wastewater because of their solubility in water-miscible solvents such as alcohols. In aqueous solution [17], they present a characteristic UV absorption (Fig. 65).

Sulphur organic compounds are very sensitive to pH (Fig. 66). As the acidic form absorbs differently from the basic one, it is possible to calculate the pK_a value of these compounds. The example of thiophenol is given, and the estimated pK_a value is close to 6.2 (Fig. 66).

3.9. Surfactants

Surfactants are the active substance of detergents always present in wastewater. Most of them are characterised by an aromatic or phenolic group in their structure and thus

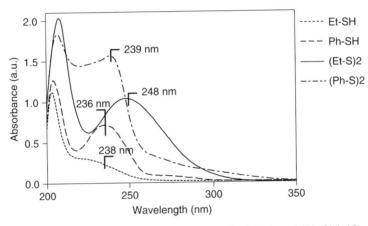

FIGURE 65. *UV spectra of ethanethiol (Et-SH) 70 mg/L, thiophenol (Ph-SH) 10 mg/L, ethyl disulphure (EtS)$_2$ 290 mg/L and phenyldisulphure (PhS)$_2$ 20 mg/L.*

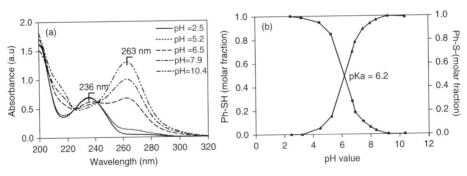

FIGURE 66. *UV spectra of thiophenol according to pH values and predominance diagram.*

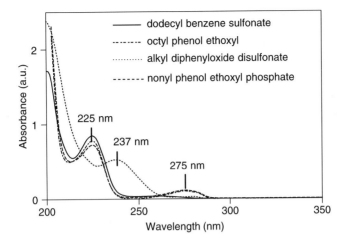

FIGURE 67. *UV spectra of aromatic surfactants.*

absorb in the UV range (Fig. 67). The global UV estimation of these compounds has been proposed considering the specific absorption of monoaromatic surfactants [18].

4. NONABSORBING ORGANIC COMPOUNDS

The absorbance of some organic substances in the UV-visible range is very weak for the concentrations usually found in water and wastewater. Some of them do not absorb, even at high concentration. However, it is possible to sometimes detect or reveal their presence with the use of a pretreatment, such as a specific reaction of derivatisation or a photo-oxidation step.

4.1. Carbonyl compounds: use of absorbing derivatives

As seen previously, some carbonyl compounds, such as formaldehyde, do not have any specific absorption band and, therefore, their spectrophotometric determination is not easy. In this case, the use of a derivative from 2,4-dinitrophenylhydrazine, namely 2,4-dinitrophenyl-hydrazone [19], is advisable since the resulting UV spectrum shows an absorption maximum at 360 nm (Fig. 68).

Aldehyde and acetone derivatives show an absorption peak around 370 nm (Fig. 69). The absorptivities of derivatives (2,4-dinitrophenylhydrazone) increase notably as compared to carbonyl compounds, making their detection more easy (Table 6).

4.2. Aliphatic amines and amino acids: photo-oxidation

Aliphatic amines, such as diethylamine, and amino acids, such as glycine and glutamic acid, do not show any specific absorption (Fig. 70). In this case, a photo-oxidation (using a low-pressure mercury lamp and a potassium peroxodisulphate solution as oxidant) could lead to nitrate formation (Fig. 71), easily determined by UV spectrophotometry (see Chapter 4).

C. Gonzalez et al.

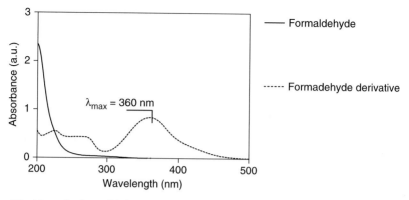

FIGURE 68. *Use of a formaldehyde derivative from 2,4-dinitrophenylhydrazone (9 mgL^{-1}).*

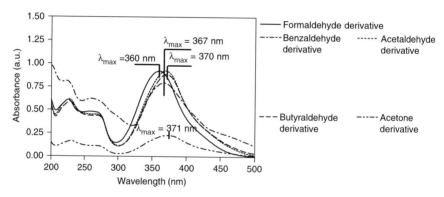

FIGURE 69. *Aldehydes and acetone derivatives from 2,4-dinitrophenylhydrazone (10 mgL^{-1}).*

TABLE 6. *Absorptivities of some carbonyl compounds and their derivatives*

Carbonyl compounds	Wavelength (nm)	ε (Lmol^{-1}cm^{-1})
Acetone	266/371*	20/5470**
Acetaldehyde	277/370	6.0/20680
Butyraldehyde	283/367	47/22290
Benzaldehyde	251/367	12571/22750
Formaldehyde	200/360	0.2/19350

*Wavelength of absorption peak of compound and derivative.
**Corresponding absorptivity values.

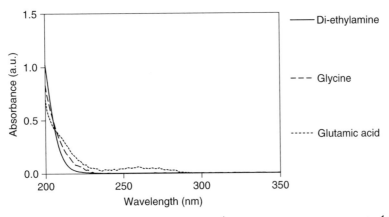

FIGURE 70. *UV spectra of diethylamine (50 mgL^{-1}) and amino acids (500 mgL^{-1}).*

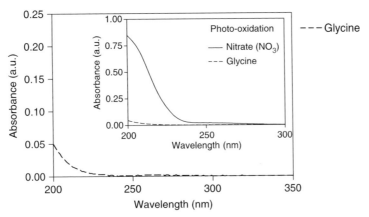

FIGURE 71. *UV spectra of glycine (27 mgL^{-1}) before and after photo-oxidation.*

In contrast with the previous case, this procedure is not specific and only show that the photo-oxidised substance contains nitrogen.

4.3. Carbohydrates: photo-degradation

Carbohydrates do not absorb in the UV-visible range, but a photo-degradation step is able to reveal their presence in an aqueous solution [20]. Contrary to the previous case where the association of UV irradiation with a chemical oxidant (peroxodisulphate) has demonstrated its efficiency, this procedure is not adapted for carbohydrate determination. The oxidation rate of this method is too fast, and absorbing intermediates (carbonyl compounds) are only weakly observed, because they are quickly oxidised into carboxylic acids. Thus, milder oxidation conditions (such as a simple UV lamp) should be used in order to provide an appropriate oxidation reaction that allows the formation of an intermediate compound with a stable absorption peak (Fig. 72).

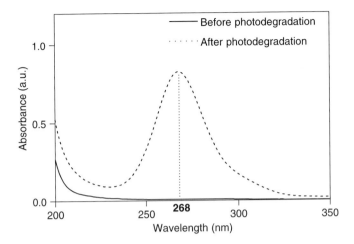

FIGURE 72. *UV spectra of sugar solution (glucose, 150 mgL^{-1}) before and after photodegradation.*

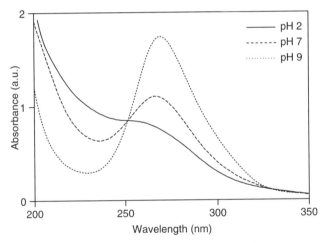

FIGURE 73. *UV spectra of sugar solution (glucose, 320 mgL^{-1}) after photodegradation: influence of pH.*

The pH value affects the UV photodegradation efficiency. From the same concentration of initial sugar solution, the absorbance obtained is greater for alkaline pH than for neutral or acidic pH. Thus, an isosbestic point is found in the pH series of the absorption spectra (Fig. 73). Alkaline conditions are needed for solutions containing low sugar concentrations. However, for solutions with high sugar concentrations, it is still better to use alkaline conditions because this reduces the analysis time.

References

1. C. Yriex, *Spécialité Chimie de l'Environnement et Santé*. PhD Thesis, Université de Provence, (1994).
2. J.M. Deroux, C. Gonzalez, P. Le Cloirec, A. Roumagnac, *The Science of the Total Environment* 203 (1997) 261.
3. N.T. Edwards, *J. Environ. Qual.*, 12 (4) (1983) 427.
4. A.N. Gennadiev, I.S. Kozin, Y.I. Pikoski, *Eurasian Soil Sci.* 30 (3) (1997) 249.
5. H.H. Perkampus, *UV–Vis Atlas of Organic Compounds*, 2nd edition, VCH, Weinheim (1992).
6. R.L.Reeves, R.S. Kaiser, *J. Org. Chem.*, (B) (1968) 1308.
7. K.M. Tawarah, H.M. Abu-Shamleh, *Dyes Pigm.*, 16 (1991) 241.
8. R.H. Peters, H. Sumner, *J. Chem. Soc.*, (1953) 2101.
9. Commission des communautés européennes, DGXII, Office des publications officielles des communautés européennes, Luxembourg (1993).
10. P.H. Howard, *Handbook of Environmental Fate and Exposure Data for Organic Chemicals*, Lewis Publishers, Michigan USA (1991).
11. S. Görög, *Ultraviolet-Visible Spectrophotometry in Pharmaceutical Analysis*, CRC Press, Inc, Boca Raton Florida 33431 (1995).
12. H.M. Pinheiro, E. Touraud, O. Thomas, *Dyes Pigm.*, 61 (2004) 121.
13. E. Cavalieri, S. Higginbothan, E.G. Rogan, *Polycycl. Arom. Comp.*, 6 (1994) 177.
14. S. Alfred, Y. Chau, B.K. Afghan, *Analysis of Pesticides in Water*, CRC Press, Inc, Boca Raton Florida, II, III (1982).
15. Des W. Connell, *Basic Concepts of Environmental Chemistry*, Lewis Publishers, Boca Raton New York (1997).
16. H. Börner, *Pesticides in Ground and Surface Water*, Springer-Verlag, Berlin Heidelberg New York (1994).
17. B. Roig, E. Chalmin, E. Touraud, O. Thomas, *Talanta*, 56 (2002) 585.
18. F. Theraulaz, L. Djellal, O. Thomas, *Tenside Surf. Det.*, 33 (1996) 6.
19. K. Blau, J. Halket, *Handbook of Derivatives for Chromatography*, John Wiley and Sons, Chichester England (1993) 170.
20. B. Roig, O. Thomas, *Analytica Chimica Acta*, 477 (2003) 325.

CHAPTER 4

Aggregate Organic Constituents

O. Thomas[a], F. Theraulaz[b]

[a]Observatoire de l'Environnement et du Développement Durable, Université de Sherbrooke, Sherbrooke, Québec, J1K 2R1, Canada; [b]Laboratoire Chimie et Environnement, Université de Provence, 3 place V. Hugo, 13331 Marseille Cedex, France

1. INTRODUCTION

Considering the available UV spectroscopic data, it seems actually very difficult to analyse specific organic compounds in water and wastewater using UV spectrophotometry, except if the concentration is high as it can be, as in the case of industrial samples (see Chapter 9). Actually, the existence of complex organic matrix limits the possibilities of organic compounds analysis, even with adapted techniques such as chromatography. This is the reason why the measurement of aggregate organic parameters (oxygen demand, for example) and constituents (phenol index, for example) is most often used for wastewater quality control.

Even if their significance is related to properties quantification (e.g. oxygen demands) or if they must be considered as tests (BOD5) or rough estimations (indexes), these parameters are essential for the knowledge of organic pollution load. In relation to specific organic determination compounds, the measurement of aggregated organic constituents brings to users a fast and global way to appreciate the potential effect of nonspecific pollution.

The main effect of organic pollution is the oxygen depletion in water. At the end of the nineteenth century, this type of chronic pollution was historically the starting point of developments for wastewater measurement and treatment. Figure 1 represents the schematic principle of biological phenomenon (aerobic), used for the characterisation and treatment of organic pollution. Organic matter can be considered as a carbon source for microorganisms present in natural water and wastewater. Obviously, the presence of oxygen and nutrients is required. The residual compounds can be simple mineral ions or molecules (if the mineralisation process is completed) or residual degradation by-products.

As oxygen depletion is the main ecological impact of organic pollution, oxygen-demand-based methods (BOD, biological, and COD, chemical) are very important.

A more complete approach is shown in Fig. 2, with details on other aggregated organics (surfactants, phenols, etc.), evolution factors (chemical and physico-chemical reactions) and residual by-products (organic acids, refractory organics, etc.).

As the measurement of BOD and COD requires 5 days and at least several hours, respectively, and considering the other limits related to their principle, the use of rapid techniques for a direct characterisation of organic matter has been developed, among which are TOC measurement and UV spectrophotometry. Notice that an initiative to use

O. Thomas, F. Theraulaz

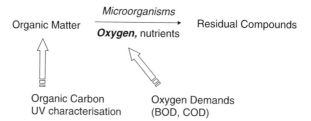

FIGURE 1. *Principle of organic matter degradation and measurement.*

a rapid-oxygen-demand technique called Total Oxygen Demand (TOD) was aborted in the 1980s due to the fragility of the instruments.

There exists a great complexity of organic compounds in natural waters as well as in effluents, from natural and/or anthropogenic origins. Most of these compounds are susceptible to be degraded by chemically or biologically, depending on the medium. For surface water, for instance, there exist natural biodegradation processes that require a long period of time. In contrast, for urban wastewater, a rapid biodegradation process is carried out inside a global treatment scheme, in a short time. However, in both cases, the global quality of water depends on the concentration of the aggregate organic constituents. This control is important both to prevent a eutrophic evolution in natural water, and to determine the elimination rate of the process, for wastewater treatment plant.

2. REFERENCE METHODS ASSISTANCE

As mentioned previously, a solution to estimate the global organic pollution of water is the measurement of the oxygen demand, which is directly correlated to organics concentration. There exist two main parameters commonly used: the biological oxygen demand (BOD) and the chemical oxygen demand (COD). BOD gives a representation of oxygen demand in the conditions of natural self-purification. BOD represents the quantity of

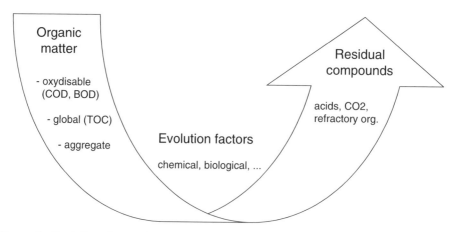

FIGURE 2. *Evolution of organic matter in water.*

oxygen consumed by micro-organisms for the oxidation of organic matter. COD uses a chemical oxidation that is more rapid, but does not concern the same oxidisable compounds as BOD. This method concerns all compounds that can be oxidised by potassium dichromate (the major part of organic compounds), and oxidisable mineral salts (sulphide, sulphite, etc.).

2.1. BOD measurement

The principle of the BOD [1] is very close to the basic principle of organic matter degradation (Fig. 1). A sample is placed in a flask, diluted with a feeding solution (nutrients) saturated in oxygen and in presence of microorganisms. The flask is placed in the dark, and the temperature is controlled at 20°C. The consumption of oxygen is followed during 5 days or more, either with a manometric device (including a CO_2 trap) or by direct oxygen measurement (for example, using a potentiometric electrode). The BOD5 value is deduced from the oxygen consumption (minus the value corresponding to a blank), taking into account the dilution factor. One important point is to be sure that the dilution is adapted to both the reagent concentration and the sample demand. A good result is obtained when the oxygen consumption is between 40 and 60% of the initial concentration (about 9 mgL^{-1} oxygen at 20°C). Thus, several dilutions (calculated from the COD value) are planned for the purpose.

The evolution of the BOD value is typically the one in Fig. 3, showing two plateaus. The first one, occurring between 3 and 6 days, corresponds to the carbonaceous degradation, and the second one, slower to appear, to the nitrification phenomenon. As the first step concerns only organic pollution characterisation, a nitrification inhibitor (allylthiourea) can be added to the flask.

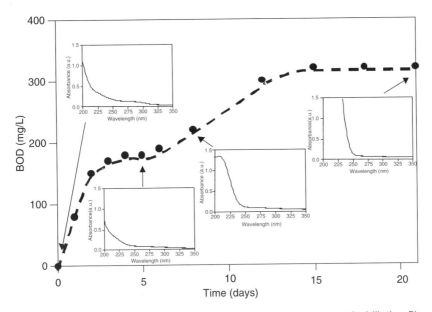

FIGURE 3. *Evolution of the UV spectrum of an urban wastewater sample (dilution 5), with BOD measurement.*

The study of UV spectrum evolution of the sample shows clearly the sequence of oxygen consumption, with a degradation of the signal related to raw organic matter (on the whole spectrum) and is followed by the high increase of nitrate response, between 200 and 230 nm (see Chapter 5), at the end of the measurement (ultimate BOD or BOD21). Thus, UV spectrophotometry avoids the use of the inhibitor by controlling the nitrification beginning.

Another use of UV spectrophotometry in BOD measurement assistance is for helping in the choice of the right dilution. As said before, the knowledge of the COD value is necessary for the dilution calculation, with the aim of a BOD value of the diluted sample around 5 mgL^{-1}. From the UV spectrum of the raw sample, it is possible to determine which dilution to apply, from a simple calibration or after BOD or COD estimation (see part 3.)

2.2. COD final determination

COD measurement is based on the use of an oxidant solution (potassium dichromate in concentrated sulphuric acid) during 2 h of mineralisation in hot acid medium. The oxidant consumption is determined by a difference with a final redox titration [1]. Since this method is rather tedious and considering the optical properties of dichromate and trivalent chromium, Cr^{3+} (this last form resulting in the oxidation of organic matter), several UV-visible spectrophotometric methods have been proposed for the final determination [2,3]. Figure 4 presents the spectra of COD test tubes solutions (range 50–1500 mg/L), corresponding to three trials (blank, industrial wastewater and concentrated ethanol solution). The dichromate visible absorption peak (at 440 nm) decreases with COD, while, in the same time, the one corresponding to Cr^{3+} increases. Compared to the corresponding spectra of hexavalent chromium (dichromate) and Cr^{3+} in neutral conditions (see Chapter 5), the peaks are more intense in sulphuric acid (hyperchromic effect), accompanied with a decrease of corresponding wavelengths (hypsochromic effect).

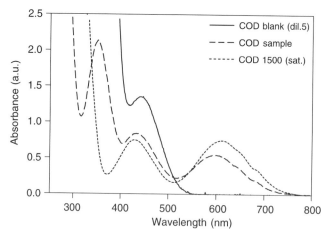

FIGURE 4. *UV-visible spectra of COD test tube solutions (blank, sample corresponding to 1107 mg/L of COD and saturated sample of 1500 mg/L).*

TABLE 1. *Absorptiometric methods for the final COD determination [3]*

COD range (mg/L)	Chromium form detected	Wavelength (nm)	$K_2Cr_2O_7$ conc.*(mol/L)
2–30	Dichromate	345 (350)	2.50×10^{-3}
10–150		440 (435)	8.33×10^{-3}
5–150		405–440 (435)–520**	8.33×10^{-3}
50–800	Cr^{3+}	610 (600)	4.17×10^{-2}
50–1500			8×10^{-2}

*In the reagent (diluted six times for the test).
**Preferable if suspended solids.

These optical properties are used for the final determination of residual hexavalent chromium or Cr^{3+} formed (Table 1) for COD determination. The choice of wavelengths (and obviously of the oxidant concentration) depends on the COD range. The final determination of COD at 610 (600) or 440 (435) nm is widely used, with filter photometers for commercial tube test methods. However, some problems in reading can exist in cases of suspended solids coming either from industrial samples, for example, or, more often, precipitation occurring during COD test. In this case, a triwavelength method can be used for interferences compensation [3] (see also Chapter 2).

2.3. TOC explanation

Total or dissolved organic carbon is the most relevant parameter for the global determination of organic pollution of water and wastewater. It has been proposed in the seventies, i.e. for automatic survey, because of the problems related to the use of BOD and COD. Actually, the organic carbon is used for that purpose or for the completion of the knowledge of organic pollution, as the other parameters quantify its main effect: oxygen consumption. It is the reason why TOC is often considered as the "true" parameter, even if its measurement is not easy [4]. Unfortunately, due to technical or metrological considerations, some problems exist with the use of TOC measurement for wastewater. TOC measurement seems to give specific information (the organic carbon content of the sample) but only leads to a global aggregate response.

The main application of UV spectrophotometry for TOC explanation is to state on the risk of global organic pollution evolution. Figures 5 and 6 present spectra of wastewater samples with the same TOC value.

The first case (Fig. 5) shows the effect of wastewater dilution leading to the same value of TOC than the treated sample. Obviously, the nature of organic matter is different as the treated effluent contains small oxygenated organic molecules, confirmed by the presence of nitrate (high Gaussian absorbance for short wavelengths; see Chapter 5). From the environmental point of view, the impact of both effluents is obviously not similar.

The second example (Fig. 6) is different. The difference between the shape of wastewater spectra is due to the distribution of suspended and colloids matters (see Chapter 6) and to the nature of dissolved organic matter. For example, the importance of the absorption peak at 225 nm (related to benzenic surfactants) of the samples is variable.

The last example is shown in Fig. 7 with two spectra, one corresponding to urban wastewater, with a mixture of organic compounds, and the other one to a phenol solution of 60 mgL^{-1}. The TOC value is the same for the two samples, but as for the first example (Fig. 5), the environmental impacts are very different. In case of the presence

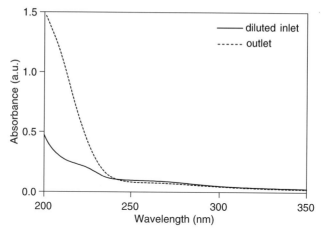

FIGURE 5. *UV spectra of raw (dilution 3) and treated (biological process) urban wastewater with the same TOC value (17.7 mgL^{-1}).*

of a major pollutant, represented here by phenol, the diagnosis for potential toxicity has to be envisaged. Chapter 9, dealing with industrial wastewater, shows several examples illustrating this problem.

The interest in the explanation of TOC is thus evident, both to state on the stability degree of organic matter in wastewater, and to estimate the complexity of the corresponding mixture, for the toxicity evaluation, for example, if a major pollutant (obviously absorbing) is present.

As the measurement of TOC is not sufficient for the determination of the biodegradable fraction of organic matter, the knowledge of which is very useful, a parameter based on the measurement of dissolved organic carbon (TOC after 1-μm filtration),

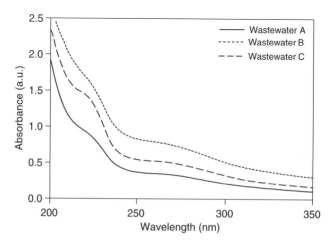

FIGURE 6. *UV spectra of three raw urban wastewater samples (TOC = 57 mgL^{-1}).*

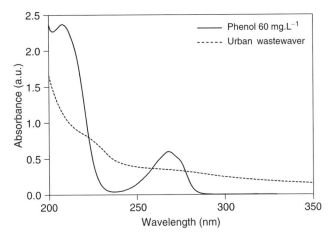

FIGURE 7. *Spectra of urban wastewater and phenol (pH 7), with the same value of TOC (60 mg/L).*

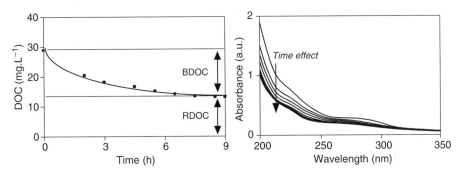

FIGURE 8. *Evolution of COD value and UV spectra during BDOC determination of an urban wastewater sample (RDOC, refractory dissolved organic carbon).*

the biodegradable dissolved organic carbon (BDOC), is sometimes used for this purpose [5,6]. Figure 8 shows how UV spectrophotometry can be helpful for BDOC determination [7].

The BDOC represents the fraction of DOC assimilated by the biomass (fixed on sand) for the conditions of the test, corresponding, in this example, to more than 50% of the initial DOC. The spectra evolution confirms the effect of the degradation process. UV spectrophotometry can be proposed to check the activated sand preparation [7], but also for a fast and simple DOC (or TOC) estimation.

3. UV ESTIMATION OF BOD, COD AND TOC

The estimation of the main aggregated organic parameters for water and wastewater is one of the first applications of UV spectrophotometry. As a lot of organic compounds absorb in the UV region (see Chapter 3), the exploitation of UV spectrum for a quick

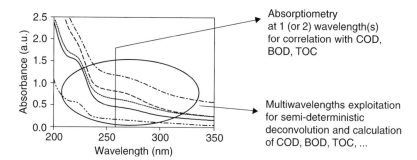

FIGURE 9. *Main approaches for aggregate parameters estimation from UV spectra exploitation.*

estimation of COD, BOD or even TOC (keeping in mind that the corresponding methods are time consuming or expensive) is interesting. Two main approaches have been proposed, depending on the number of wavelengths to be considered (Fig. 9).

The first way is very simple as it is the research of a "useful" correlation between one wavelength absorbance measurement and the corresponding parameter value. This simple absorptiometry has been carried out, without actual success, from the use of a low-pressure mercury lamp emitting principally at 254 nm [8–11]. This method was proposed more for economic than scientific reasons, (cheap UV light source). The use of a second wavelength (in the visible range) for interference compensation (due to suspended solids) does not improve the quality of results.

Without getting into details of the different studies related to the "history" of UV spectra exploitation and application (see Chapter 2 for methods and Chapters 7 to 10 for applications), it is obvious that multiwavelength exploitation methods are more robust. Some works propose the use of PLS procedure for COD estimation [12], but the spectra deconvolution with the use of a semi-deterministic procedure is more powerful [13,14]. The mathematical principle of the procedure has been explained in Chapter 2, but its application for aggregated organic parameter estimation is presented hereafter. Two main steps are carried out during the procedure. The first one is the mathematical deconvolution (modelling) with a given basis of reference spectra, completed by the checking of the restitution quality with the study of the error between the actual spectrum and the restituted one after deconvolution. The second step is the parameter calculation from a calibration file.

3.1. UV spectra modelling

The use of UV spectrophotometry for organic pollution characterisation and estimation is obvious because of the presence of a lot of absorbing compounds in water and wastewater. Even if it is quite possible to find some nonabsorbing organic molecules (saturated hydrocarbons, small molecules, sugars, etc.), a UV spectrum of water and wastewater is very rarely flat.

Starting from the study of several thousands of urban wastewater samples, a first "universal" basis of reference spectra has been proposed for one application concerning the wastewater quality monitoring [12,13]. The selected reference spectra are displayed in Fig. 10.

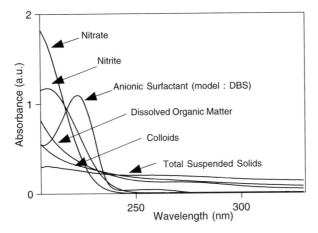

FIGURE 10. *Normalised reference spectra of the "universal basis" of the semi-deterministic method (DBS: dodecylbenzenesulphonate).*

The choice of reference spectra is carried out, on one hand, "manually" from real (deterministic) spectra, with the consideration of determined spectra (nitrate, nitrite, surfactants). It is completed automatically, on the other hand, with a mathematical procedure [15] allowing the selection of the more relevant spectra for the model, able to explain by a linear combination, the shape of UV spectra of water or wastewater. This last procedure can be replaced by any advanced statistical algorithms (PCA or PLS, for example) or by commercial software (such as UVPro® from Secomam).

This semi-deterministic approach is interesting because the basis of reference spectra used for the modelling of real spectra is made up of spectra mixtures and specific mineral or organic compounds, the optical properties of which often explain a part of the UV spectra. The reference spectra of mixtures are statistically representative of the different heterogeneous fractions of wastewater, because they are selected from wastewater fractionation. For each wastewater sample, several filtrations are carried out (1 and 0.025 μm), and the spectra of the filtrate are acquired. The differential spectra are then considered for the basis constitution (Table 2).

This procedure can be generalised for the study of surface or seawater, for example. In these cases, some reference spectra (suspended solids, colloids, surfactants) are replaced by more relevant spectra for the medium, such as humic substances, mineral-suspended solids or chloride) in order to constitute specific but rather "universal" basis of reference.

TABLE 2. *Origin of reference spectra of mixtures (see Fig. 10)*

Name of reference spectrum	Origin (urban wastewater)
Total suspended solids	Difference between spectra of raw and filtered (1 μm) sample
Colloids	Difference between filtered samples (at 1 and 0.025 μm)
Dissolved organic matter	Spectrum of filtrate (at 0.025 μm)

3.2. Parameter calculation and calibration

After the deconvolution step, giving the contribution coefficients of reference spectra (see Chapter 2), the parameters calculation is possible by using the same coefficients and a corresponding calibration file (Fig. 11). This calibration file includes the corresponding concentrations for specific compounds (nitrate, nitrite, anionic surfactants, etc.) and the values related to the reference spectra of mixtures (Table 3). The latter are statistically calculated for the purpose, through a preliminary stepwise regression study, from a set of at least 30 samples (with 30 corresponding values of parameters and 30 sets of contribution coefficients).

For each basis of reference spectra, the choice of which is guided either by the knowledge of the sample origin, or the restitution error value, a corresponding calibration file

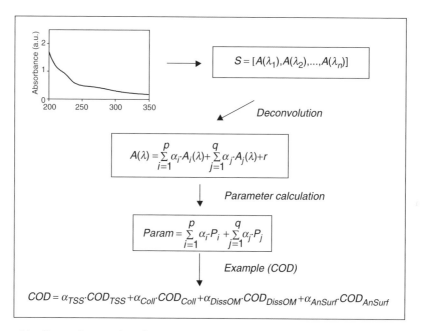

FIGURE 11. *General procedure for aggregate organic parameter estimation (e.g. chemical oxygen demand) with a model of p reference spectra of mixture and q of specific compounds (r is the restitution error, which must be minimal; see Chapter 2).*

TABLE 3. *Reference spectra contribution for urban wastewater calibration file*

Ref. spectra	COD	BOD	TOC	Surfactants	Nitrate	TSS
TSS	✓	✓	✓	—	—	✓
Colloids	✓	✓	✓	—	—	—
Dissolved OM	✓	✓	✓	—	—	—
Surfactants	✓	✓	✓	✓	—	—
Nitrate	—	—	—	—	✓	—

(For surfactants, see next section; for nitrate and TSS, see Chapters 5 and 6.)

has to be used. Actually, a few models (basis of reference spectra + calibration file) are sufficient for the main applications concerning water and wastewater quality monitoring.

3.3. Validation

Several validation experiments have been carried out for water and wastewater quality monitoring. Tables 4 and 5 present the general conditions and the results of UV estimation of aggregated organic parameters.

The proposed UV method has been tested from a lot of real samples corresponding to different treatment plants and rivers or lakes for each water type (at least 30). The results have been compared with the ones obtained with reference methods, chosen from among international standards. The comparison is carried out for each parameter by calculating the determination coefficient (R^2) and the parameters of the regression line (slope and intercept). The adjustment between the measured and estimated values of aggregate parameters is quite satisfactory, with a slope and intercept close to 1 and 0, respectively. Figure 12 presents an example of graphical results for the estimation of BOD5 of natural water.

The study of numerical results related to the regression study for each aggregated parameter comparison is not sufficient. On one hand, the distribution of the results must be carefully checked, for example, with the study of the residuals variation (Fig. 13), corresponding to a residual standard deviation (RSD) of 20%. In this case, no particular

TABLE 4. *Validation experiments for UV estimation of aggregate organic parameters (generalities)*

Parameter (unit)	Reference method	Samples	Number	Observation
BOD5	Dilution	Urban wastewater	120	Raw and treated
		Natural water	55	Surface water
COD	Potassium dichromate	Urban wastewater	140	Raw and treated
		Natural water	55	Surface water
TOC	High temperature	Urban wastewater	75	Raw and treated
		Natural water	40	Surface water
DOC	High temperature	Urban wastewater	75	Raw and treated

TABLE 5. *Validation experiments for UV estimation of aggregate organic parameters (results)*

Parameter	Water*	Range	Determination coefficient	Slope	Intercept
BOD5	Urban ww.	5–250 mg O_2 L^{-1}	0.89	0.84	22.6
	Natural w.	1–15 mg O_2 L^{-1}	0.92	0.94	0.7
COD	Urban ww.	10–500 mg O_2 L^{-1}	0.91	1.05	2.0
	Natural w.	2–50 mg O_2 L^{-1}	0.93	0.95	0.6
TOC	Urban ww.	5–150 mg C L^{-1}	0.96	0.99	2.5
	Natural w.	0.5–10 mg C L^{-1}	0.93	0.94	0.1
DOC	Urban ww.	5–150 mg C L^{-1}	0.98	0.98	1.6

*ww.: wastewater, w.: water.

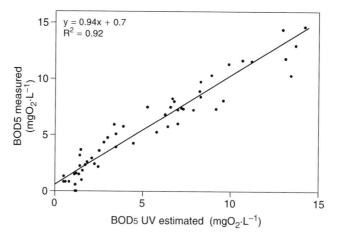

FIGURE 12. *Regression line between estimated and measured BOD5 for natural water.*

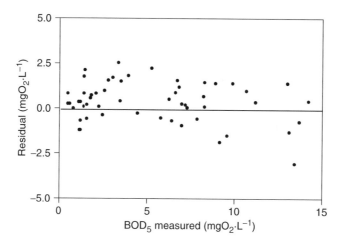

FIGURE 13. *BOD5 residuals distribution (measured – UV estimated values).*

distribution shape is noticed, indicating that a linear regression model can be envisaged for the estimation. On the other hand, a statistical comparison can be used [16] in order to test the validity of the linear model (slope and intercept not different from 1 and 0). For the previous results, all validation experiments have led to the acceptance of the UV procedure as an alternative method for the estimation of the corresponding aggregated parameter, except for the BOD5 estimation of urban wastewater.

4. UV ESTIMATION OF CLASS OF ORGANIC COMPOUNDS

4.1. Surfactants (anionic)

Surfactants are used as cleaning agents for domestic or industrial applications. Commonly, these molecules are made up of two parts with very different characteristics.

The long hydrocarbon chain forms the hydrophobic tail, and the polar group (carboxylic, sulphate, phosphate, ammonium, ethoxylate) forms the hydrophilic head. The tail group would rather dissolve in nonpolar material such as grease, whereas the head group has a great solubility in water. Therefore, surfactants are oriented towards interfaces such as oil–water and form a surfactant monolayer that greatly reduces the surface tension.

Surfactants are classified into four categories according to the nature of the charge of the hydrophilic part of the molecule:

- anionic surfactants
- cationic surfactants
- nonionic surfactants
- amphoteric surfactants

The anionic category represents the major part of surfactant consumption. Anionic surfactants include linear alkylbenzene sulphonates (LASs), alcohol sulphates or alcohol ethersulphates. The latter compounds have a poor absorption in the UV range, whereas the LASs show a significant absorption (see Chapter 11). Dodecylbenzene sulphonate (DBS) is used as a reference for LAS measurements (standard method) and shows an absorption band at 225 nm (Fig. 14).

Other anionic surfactants, such as nonyl phenol ethoxy phosphate (commercial term) with an aromatic structure, also absorbs in the UV range at 225 nm. Nonyl phenol ethoxy phosphate has two absorption bands (225 nm, 275 nm). The presence of two aromatic rings in the molecular structure of alkyl diphenyloxide disulphonate (commercial term) leads to a bathochromic shift due to an extended conjugation in molecule. In this case, the absorption band is observed at 237 nm.

Nonionic surfactants such as polyethylene glycol *p*-isooctylphenyl ester (octoxynol-9) also present two absorption maxima (225 nm, 275 nm). Fatty alcohol ethoxylates do not absorb in the UV region.

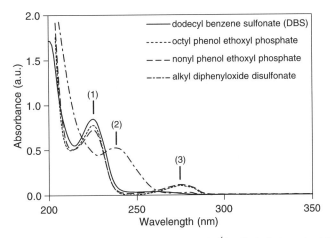

FIGURE 14. *Surfactant spectrum of octynol-9 (50 mgL^{-1}), alkyl diphenyloxide disulphonate (50 mgL^{-1}), nonyl phenol ethoxy phosphate (50 mgL^{-1}) and DBS (30 mgL^{-1}).*

Cationic surfactants (quaternary ammonium group) and amphoteric surfactants with a long alkyl chain present a poor absorption in UV.

Among all aggregate organic constituents that can be found in natural and urban wastewaters, the major ones are certainly surfactants. For several decades, there has effectively been an important increase in the domestic use of detergents in the entire world. This phenomenon has led to surfactants becoming one of the major parameters of water pollution today. More particularly, the survey of urban wastewater discharges from treatment plants into rivers or lakes is now a necessity. We must take into account the important visual impact due to the possible appearance of foam downstream of the discharge [17], and also of some toxic effects depending on surfactant concentration (if higher than 0.5 mg/L) [18]. It is the principal reason why the remaining surfactants responsible for this phenomenon are often used as a fingerprint of urban pollution. We can easily consider anionic surfactants, and more specifically the LAS family, as the most common class of surfactants existing in wastewater [19], and as the most cost-effective products in widespread commercial use.

Subsequently, there exists a great interest in the rapid determination of this group of compounds in natural and wastewaters, but there is today an important lack of a routine analysis. Actually, the most common procedure, called MBAS (methylene blue active substances), is nonspecific and time consuming due to the extraction step, and is sensitive to many interferences. For more precise determination of LAS, there exist a lot of methods such as GC-MS or HPLC techniques, which are very relevant [20,21], but equally sophisticated and totally unadapted to a simple and rapid determination. Even the simple UV absorptiometric methods using absorbance measurement at few wavelength (1 to 3) [22] or the derivative signal of the spectrum [23] have their limitation (existence of a spectral background, poor quantification limit, nonlinear interference signal, etc …) and are not relevant for all types of samples.

The solution to this problem is to use the association of the direct UV spectrophotometry and of the semi-deterministic deconvolution of the absorbance spectrum, which is presented in Chapter 3. In fact, the existence of a shoulder in the UV absorption spectrum (near 220–225 nm), commonly seen for the major part of urban wastewater samples, is obviously associated with the existence of highly absorbing compounds such as LAS [12,24].

Starting from the recent development of the UV multiwavelength deconvolution method designed for natural and wastewater examination [12] previously described (Chapter 2), several defined reference spectra, and different combinations of them, lead actually to the constitution of three basis of reference spectra, used for the restitution of spectra of natural and urban wastewater samples [25]. These three bases allow restituting, respectively, raw or biologically treated sewage, physico-chemically treated sewage, and natural water. In this approach, the type of reference spectra is not all related to specific compounds. Notice that the spectrum related to LAS appears in each basis, as these compounds can be present in natural water or wastewater.

Such a method has taken into account that LAS are a mixture of various alkyl homologues (ranging from C10 to C14) and phenyl positional isomers, the composition of which varies from the commercial products to the wastewater treatment.

This procedure has been validated on different samples of wastewater (inlets and outlets of biological and physico-chemical wastewater treatment plants) to which standard amounts of DBS has been added (Fig. 15).

The standard addition method is used taking into account the poor precision of the reference procedure (MBAS), and the concentration of the added LAS was included between 0 to 25 mg/L, and was chosen in agreement with the common values encountered in

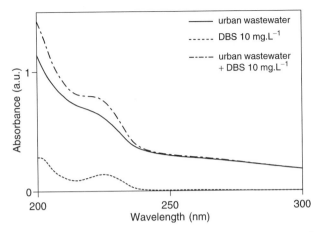

FIGURE 15. *Spectra of DBS (dodecylbenzenesulphonate) solution and urban wastewater with and without standard addition.*

such samples. The values of the recovered addition of DBS, calculated as the difference between the determined concentration after addition (i.e. taking into account initial and added detergents) and the initial one, which are calculated with UV method, were compared with the real added concentration. The relative error calculated between the measured and theoretical values shows that the results are very close. Figure 16 shows the regression line resulting from the comparison with a very good determination coefficient ($R^2 = 0.99$).

Analytical characteristics of the proposed method are as follows. The detection limit (0.20 mg/L) is given as three times the standard error value for the blank, chosen as a biologically treated wastewater sample with negligible detergent concentration. The sensitivity is 0.04 mg/L, and the analytical range depends on the spectrum saturation occurring

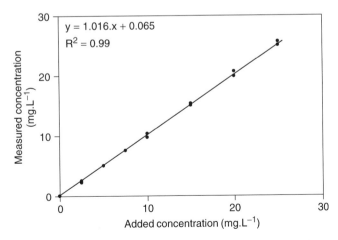

FIGURE 16. *Regression line resulting from the comparison between estimated and added DBS concentration in a water sample.*

for an absorbance value of about 2.5 a.u., taking into account that other components absorb (essentially the organic matter); the maximal concentration is therefore about 50 mg/L for a 10-mm optical path length, without sample dilution. The precision is calculated for a concentration of 15 mg/L of DBS in a sewage sample. Its value (2.4%) shows that this rapid method is more precise than the MBAS determination (RSD between 9 and 14%) [1].

4.2. Phenol index

Phenolic compounds, i.e. hydroxyl aromatic compounds, are widely used and produced by industries. Due to the toxicity of some of them, their discharge (wastewater) or their storage (industrial waste) must be controlled. The phenol family groups the phenol molecule and several substituted compounds, such as chlorinated phenols (or halogenated), cresols (with methyl group), amino or nitro phenols, the chemical properties of which are different. Considering that the analytical determination of phenolic compounds needs different chromatographic procedures [1], including some cleaning steps for interferences removal and preconcentration, an aggregated method called "phenol index" is often used for aqueous effluents [1]. This method consists of two main steps. A steam distillation is carried out in acidic medium, and then a colorimetric measurement is used with amino-4 antipyrine as reagent. This analytical procedure is time consuming and not quantitative, since some phenolic compounds as nitrophenols, aminophenols or polyphenols (catechols) do not fully react.

According to their structure (aromatic ring), these compounds are easily detected by UV spectrophotometry (as, for example, in HPLC procedures). Moreover, a characteristic bathochromic shift in basic medium can also be used for their determination (see Chapter 4). On the other hand, UV spectrum exploitation can be improved with the use of a second derivative [26] or the semi-deterministic method already presented. As illustrated in Fig. 17, the second derivative of UV spectra corresponding to a standard

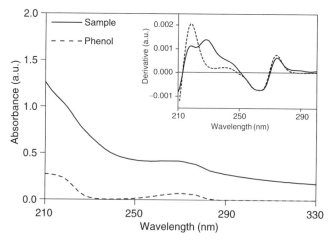

FIGURE 17. *Spectra and second derivative of phenol standard solution and wastewater sample UV spectra (phenol concentration close to 5 mg/L^{-1}).*

phenol solution and a spiked wastewater sample shows a good similarity to the phenol-specific wavelength (close to 270 nm). However, this method needs to take into account the target phenolic compound and to have a zero value for the corresponding second derivative value of interferences matrix. Figure 18 shows some examples of chemical- and petrochemical-wastewater-containing phenols.

The semi-deterministic method already presented, based on the deconvolution of the UV spectrum, can be used as an alternative method for the rapid estimation of the "phenol index". UV spectra are acquired from raw samples and, after a pH correction (pH = 12), in order to exploit both the bathochromic and hyperchromic effects, observed for the dissociated forms of phenolic compounds (phenates). This well-known procedure is particularly useful for nitrophenols, the spectra of which present a specific adsorption between 360 and 500 nm (Fig. 19). On the other hand, phenol and 2-chlorophenol have their adsorption peak range from 250 to 360 nm. It should be noticed that below 250 nm,

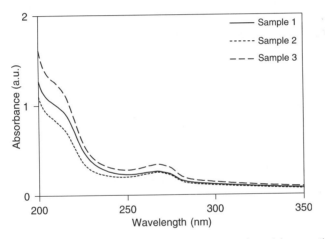

FIGURE 18. *UV spectra of industrial-wastewater-containing phenol (respectively 14.2, 16.6 and 18.2 mg/L^{-1}).*

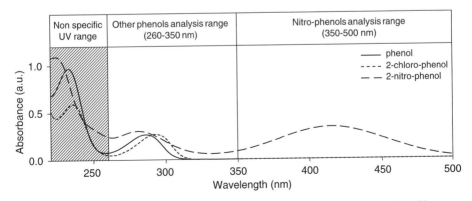

FIGURE 19. *Specific UV-visible absorption range of phenolic compounds (pH 12).*

the adsorption band is not specific to phenolic compounds, because a lot of organic compounds have an absorption in this UV spectral window.

The proposed method for the quantification of phenolic compounds includes mainly two deconvolution steps (Fig. 20). The first step is carried out for a spectral window range of 360 nm and 500 nm and allows the quantification of nitrophenol compounds (C1). Then the second deconvolution is performed on the residual spectrum for a spectral window range of 250 nm and 360 nm and leads to estimating the concentration of the other phenolic compounds (C2).

The total concentration (C1 + C2) can be considered as a good estimation of global phenolic compounds concentration. According to the principle of the deconvolution method, the difference between the sample spectrum and the restituted one is defined by the quadratic error (see Chapter 3). If the quadratic error is higher than 5%, it means that other compounds are present and could interfere with the estimation of phenolic compound concentrations. In this case, no quantification is possible but a qualitative result can be displayed (presence of unknown compounds).

A comparison between expected concentrations (from ultra-pure water spiked with phenol mixtures) and UV-visible estimated concentrations shows a good agreement (Fig. 21). Phenolic compounds used for the purpose are listed in Table 6 and were

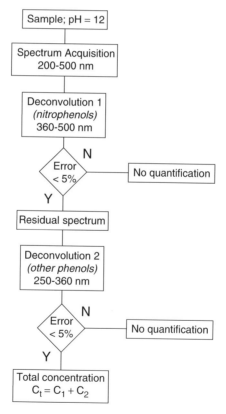

FIGURE 20. *Deconvolution procedure for the estimation of the concentration of global phenolic compounds.*

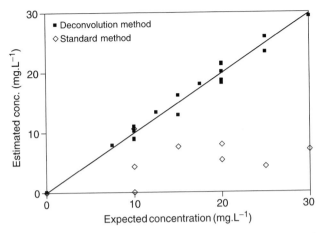

FIGURE 21. *Comparison between expected concentrations and estimated concentrations by UV-visible (deconvolution) method ($R^2 = 0.98$) and standard method ("phenol index").*

TABLE 6. *Selected phenolic compounds*

Nonsubstituted phenols	Methyl-phenols	Chloro-phenols	Nitro-phenols
Phenol	4-cresol	2-chlorophenol	2-nitrophenol
Naphtol	2,4-dimethylphenol	4-chlorophenol	4-nitrophenol
		2,4-dichlorophenol	2,4-dinitrophenol
		Trichlorophenol	
		Pentachlorophenol	

selected according to EPA priority pollutants list. The same comparison with "phenol index" measurement points out the limits of the standard method.

In order to check the potential interference of aqueous matrix, an industrial wastewater sample was spiked with phenol mixtures (Fig. 22). The correlation between concentrations estimated by the proposed UV-visible method and expected concentrations again gives satisfactory results and shows the presence of phenolic compounds in the industrial wastewater sample (around 15 mg/L).

In conclusion, the deconvolution procedure seems to be a suitable method for the rapid and simple estimation of global phenolic compound concentration. Commercial built-in spectrophotometer software such as UVPro® from Secomam or any multicomponent procedure can be used for this purpose. In this last case, samples have to be filtered in order to avoid interference due to the high TSS level in UV spectrophotometry measurement. Nevertheless, this method must be extended to other phenolic families such as aminophenols or polyphenols.

4.3. PAH (index)

Soil contamination is one of the main environmental problems, mainly in industrial countries, and is very often linked to water resource degradation. The diagnosis of potentially

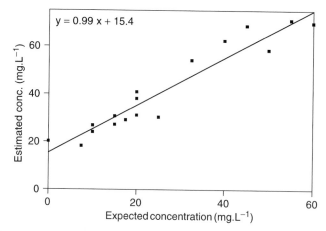

F<small>IGURE</small> 22. *Correlation between expected and estimated concentrations for spiked wastewater sample (R² = 0.91).*

polluted sites is often difficult due to the lack of simple procedures, and the treatment scheme must be defined and carried out with respect to soil characteristics and future use. The classical approach includes sampling and laboratory analysis, generally with HPLC or GC analysis of organic extracts, leading to a good selectivity [27–29].

Initial diagnosis of contaminated soils requires getting a rapid estimation of the nature and extent of pollution, and the simple evaluation of global PAH concentration is sufficient for guiding the treatment choice. According to their aromatic structure, PAH absorb strongly in the UV range (see Chapter 3), and a PAH UV index has been developed. It is based on the UV spectrophotometric analysis of the soil organic extract and gives a global PAH estimation in reference to the 16 USEPA PAH (see Chapter 3). The extraction step is carried out with acetonitrile [30], and can be improved with a solid-phase extraction (SPE) step [31].

The soil organic extract corresponding to a concentrated and purified soil PAH solution is obtained through the simple procedure described in Fig. 23.

This procedure has been applied to nearly 80 samples of soils from various industrial origins. It leads to define two types of UV organic extracts spectra presented in Fig. 24. Each UV spectrum shows a structured shape with high absorbance at the beginning of the spectrum, which decreases after 300 nm. It can be noticed that absorbance value over 350 nm is more important for the type 2, according to the bathochromic effect observed for heavy PAHs. Indeed, it has been shown that UV spectrum of type 1 corresponds to soils mainly contaminated by light PAHs (2 or 3 cycles) and UV spectrum of type 2 to soils mainly contaminated by heavy PAHs (4 or more cycles) [32].

Two specific peaks, located respectively at 254 and 288 nm, are always present on UV spectra profiles. The first one is characteristic of the presence of the 16 USEPA PAHs (Chapter 3). Thus, for a quantitative application, the absorbance value at this wavelength measurement is proposed as a PAH index for a simple estimation of global PAH concentration in contaminated soil. A validation of this approach is given by HPLC analysis of the 16 USEPA PAHs (Fig. 25).

Dealing with the specific absorbance at 288 nm, it has been observed, from the study of UV organic extracts spectra, that the ratio between the absorbance values at 254 and

FIGURE 23. *PAH-contaminated soils: diagnosis procedure.*

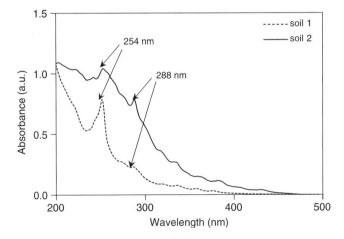

FIGURE 24. *PAH-contaminated soils extract UV spectra.*

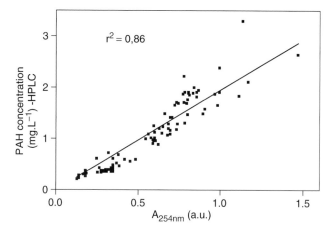

FIGURE 25. *Correlation between HPLC measurement and absorbance value at 254 nm.*

288 nm, $A_{254\ nm}/A_{288\ nm}$, varies from one soil to another (between 1 and 4) and appears to be significant of the proportion of light and heavy PAHs in contaminated soil. By the way, a rather good correlation has been found between this ratio value and the relative proportion of three-cycles PAH (light PAHs), measured by HPLC (Fig. 26). It must be specified that naphthalene (two cycles) has never been found in the studied PAH-contaminated soils.

This simple method of UV spectra exploitation (mono wavelength correlation) has been used and leads to the definition of a PAH index suitable for a rapid diagnosis of PAH-contaminated soils. Moreover, the absorbance value ratio of the two main characteristic peaks gives information about the PAH distribution in terms of light and heavy PAHs. These tools appear to be relevant with regard to the management of contaminated soils.

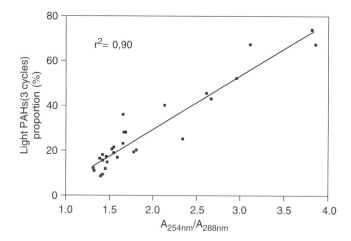

FIGURE 26. *Correlation between* $A_{254\ nm}/A_{288\ nm}$ *ratio and three cycles PAH presence.*

4.4. Other classes of organic compounds

4.4.1. Aromatic amines

As with all aromatic compounds, aromatic amines absorb relatively strongly in the UV region (see Chapter 11). A first study [33], based on the use of the deconvolution method, has been applied to the estimation of aniline derivative concentrations in industrial wastewater. For the purpose, a basis of reference spectra (see Chapter 2) has been defined by including characteristic average spectra for global and chlorinated aniline mixtures.

Another study on aromatic amines from azo dye reduction gives a complete overview on the use of direct UV spectrophotometric detection in textile industry wastewater [34].

4.4.2. Mercaptans

Mercaptans (R–SH) [35] are weak acids (pKa around 10) and give rise to an equilibrium in aqueous solution. In basic media, bathochromic and hyperchromic shifts are observed in all cases. This effect is particularly marked with thiophenate ion because of a stabilisation of the negative charge with the π electrons of the aromatic ring. By adjusting the pH of sample to 11 after addition of sodium hydroxide solution 2.5 M, the spectra show a well-defined peak of absorbance at 238 nm for the alkylthiols or 263 nm for the thiophenols. The application of the deconvolution method allows the estimation of the global concentration of mercaptans in wastewater.

A major point is related to the need to adjust the sample pH to 11. At this value, all mercaptans are supposed to be under the thiolate form, which is known to be oxidised readily at high pH value in the presence of dissolved oxygen. Thus, the spectra acquisition must be carried out less than 15 min after pH adjustment.

4.4.3. Global sugars

Sugars (carbohydrates) are non-UV-absorbing molecules, but their UV measurement is possible, for high content of sugar, by using a UV/UV procedure [36]. Firstly, after addition of a pH 9.0 buffer solution, the sample is UV-irradiated (for example, with the same system as for N compounds; see Chapter 5) for 10 min. Under the influence of UV radiation, sugars are oxidised to UV-absorbing carbonyl compounds characterised by a maximum absorbance at 268 nm. The formation of these compounds may be monitored by UV absorption spectrophotometry at the wavelength of maximum absorbance from the whole spectrum using the deconvolution method. Contrary to N compounds, the use of a chemical oxidant (peroxodisulphate) is not needed because carbonyl compounds are too quickly oxidised.

The UV/UV method allows the simple determination of overall sugar content. The procedure was applied successfully to food liquid products (soft drinks and commercial and natural fruit juices), with high sugar concentrations (between 40 and 500 g/L). The method could be extended to the determination of sugar (at lower concentrations) in industrial processes, as in wine production, for example.

5. UV RECOVERY OF ORGANIC POLLUTION PARAMETERS

At the end of this chapter, it can be interesting to qualitatively compare the different parameters with the UV response of families of organic compounds. Figure 27 presents different domains (families of organic and mineral compounds) that are related to the aggregate organic parameters and to the UV samples response.

The common part of UV and classical parameters (TOC, COD and BOD) is the biodegradable fraction of organic matter. The most comparable parameter is certainly the TOC, which also includes carbohydrates (sugars) and aliphatic (saturated) hydrocarbons not absorbed in the UV region. Some specific compounds as nitrates are associated with the UV response.

More precisely, the main organic compound families are more or less recovered by the aggregate parameters and by UV spectrophotometry (Table 7). Additional comments can be made concerning organic compounds containing hetero-atoms (N or S). They are at

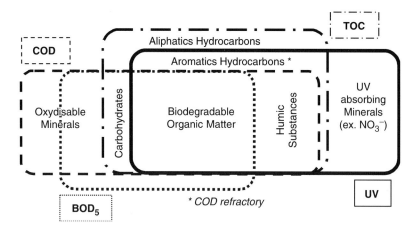

FIGURE 27. *Significance of aggregate organic parameters and UV response.*

TABLE 7. *COD, BOD, TOC and UV responses of organic compounds*

Compounds	COD	BOD	TOC	UV
Saturated compounds	P	P	Y	N
Aliphatic unsaturated hydrocarbons	Y	Y	Y	P
Aromatic compounds	P	N	Y	Y
Acids	Y	P	Y	P
Aldehydes, ketones	P	P		P
Alcohols	Y	P	Y	N
Phenolic compounds	Y	P	Y	Y
Aliphatic amines	P	P	P	N
Aromatic amines	P	P	Y	Y
N unsaturated heterocycles	N	N	Y	Y
S unsaturated heterocycles	P	N	P	Y
Humic-like substances	P	P	P	Y

Y: 90–100% of conversion or high absorption or absorption after photo-oxidation.
P: partially converted or some absorbing compounds.
N: nonabsorbing compounds.

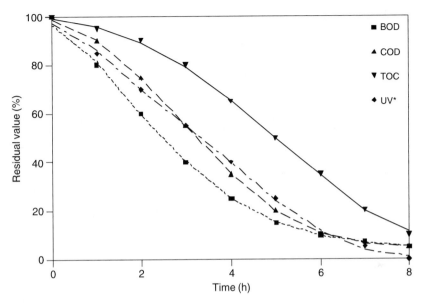

FIGURE 28. *Theoretical evolution of aggregate organic parameter and UV response with degradation time of urban wastewater (UV* expressed, for example, as the area under spectrum for wavelength >230 nm).*

best partially oxidised by COD or BOD, or totally for aromatic amines and N heterocycles by TOC. For these compounds, UV spectrophotometry is very relevant.

If we consider the variation of these parameters during a biodegradation process (Fig. 28), the evolution is very close, with some differences, however. The BOD decrease is obviously faster than for the other parameter, and TOC presents, at the beginning, a latent period related to the production of by-products containing organic carbon. The UV decrease response is very close to the one of COD but tends to zero towards the end, even if some small end-products are present, like simple carboxylic acids. The latter absorbs in the far UV region (<200 nm), showing only the tail of their spectra when their concentrations are high.

References

1. APHA, AWWA, WEF, *Standard Methods for the Examination of Water and Wastewater*, 18th edition, A.E. Greenberg, L.S. Clesceri, A.D. Eaton (Eds.), NW Washington (1992).
2. O. Thomas, Y. Munginda, *T.S.M. l'Eau*, 6 (1980) 277.
3. O. Thomas, N. Mazas, *Analusis*, 14 (1986) 100.
4. O. Thomas, H. El Khorassani, E. Touraud, H. Bitar, *Talanta*, 50 (1999) 743.
5. J-C. Joret, Y. Levi, *Trib. Cebedeau*, 39 (1986) 3.
6. J. Frias, F. Ribas, F. Lucena, *Water Res.*, 29 (1995) 2785.
7. O. Thomas, N. Mazas, C. Massiani, *Envir. Technol.*, 14 (1993) 487.
8. R.C. Hoater, *Wat. Treat. Exam.*, 2 (1952) 9.
9. R.A. Dobbs, R.H. Wise, R.B. Dean, *Water Res.*, 6 (1972) 1171.
10. B. MacCraith, K.T.V. Grattan, D. Connolly, R. Briggs, W.J.O. Boyle, M. Avis, *Sens. Actuators B*, 22 (1994) 149.

11. M. Mrkva, *Water. Res.*, 9 (1975) 587.
12. G. Langergraber, N. Fleishmann, F. Hofstadter, *Wat. Sci. Technol.*, 47 (2003) 63.
13. O. Thomas, F. Théraulaz, M. Domeizel, C. Massiani, *Envir. Technol.*, 14 (1993) 1187.
14. O. Thomas, F. Théraulaz, C. Agnel, S. Suriany, *Envir. Technol.*, 17 (1996) 251.
15. S. Gallot, O. Thomas, *Fresenius J. Anal. Chem.*, 346 (1993) 976.
16. Afnor, Experimental standard, XPT90-210, Paris, 1998.
17. W.A. Sweeney, R.G. Anderson, *J. Am. Oil Chem. Soc.*, 66 (1989) 1844.
18. M.A. Lewis, *Water Res.*, 25 (1991) 101.
19. A. Marcomini, W. Giger, *Tenside Surf. Det.*, 25 (1988) 226.
20. Q.W. Osburn, *J. Am. Oil Chem. Soc.*, 63 (1986) 257.
21. R.D. Swisher, *Surfactants Biodegradation*, Vol. 18, 2nd edition, Marcel Dekker, New York (1987).
22. D.M. Gabriel, *in Anionic Surfactants: Chemical Analysis*, J. Cross ed., Marcel Dekker, New York (1977).
23. H. Hellmann, *Tenside Surf. Det.*, 31 (1994) 200.
24. S. Suryani, F. Theraulaz, O. Thomas, *Trends Anal. Chem.*, 14 (1995) 457.
25. F. Theraulaz, L. Djellal, O. Thomas, *Tenside Surf. Det.*, 33 (1996) 6.
26. A.R. Hawthorne, S.A. Morris, R.L. Moody, R.B. Gammage, *J. Envir. Sci. Health*, A19 (1984) 253.
27. ISO standard 13877, Soil quality, PAH determination, 1995.
28. I. Baranowska, W. Szeja, P. Wadilewski, *J. Planar Chromat.*, 7 (1994) 137.
29. USEPA, Method 8310, Polynuclear aromatic hydrocarbons (high liquid chromatographic method), Method 8100, Polynuclear aromatic hydrocarbons, 1986.
30. E. Touraud, M. Crône, O. Thomas, *Field Anal. Chem. Technol.*, 2 (1998) 221.
31. O. Cloarec, C. Gonzalez, E. Touraud, O. Thomas, *Anal. Chim. Acta*, 453 (2002) 245.
32. M. Crône, *PAH Contaminated Soils Diagnosis with UV Spectrophotometry*. PhD thesis, University of Pau et des Pays de l'Adour, France, (2000).
33. F. Perez, *Spectrophotometric Study of Industrial Effluents – Application in Parameters Estimation*. PhD thesis, University of Aix-Marseille; II, (2001).
34. H.M. Pinheiro, E. Touraud, O. Thomas, *Dyes Pigm.*, 61 (2004) 121.
35. E. Chalmin, B. Roig, O. Thomas, *Talanta*, 56 (2002) 585.
36. B. Roig, O. Thomas, *Anal. Chim. Acta*, 477 (2003) 425.

UV-Visible Spectrophotometry of Water and Wastewater
O. Thomas and C. Burgess (Eds.)

CHAPTER 5

Mineral Constituents

B. Roig[a], F. Theraulaz[b], O. Thomas[c]

[a]Laboratoire Génie de l'Environnement Industriel, Ecole des Mines d'Alès, 6 Avenue de Clavières,
30319 Alès Cedex, France; [b]Laboratoire Chimie et Environnement, Université de Provence,
3 place V. Hugo, 13331 Marseille Cedex, France; [c]Observatoire de l'Environnement et du
Développement Durable, Université de Sherbrooke, Sherbrooke,
Québec, J1K 2R1, Canada

1. INTRODUCTION

The previous sections have concerned several organic compounds and their UV-visible spectrophotometric responses. Some inorganic constituents can also be studied directly with this technique, such as several oxyanions, for example. Figure 1 presents the different species studied.

Among the different mineral constituents potentially dissolved in water, several groups can be considered with regard to their environmental interest and nature:

- major minerals coming from the geochemical history of water and including cations (Na^+, K^+, Ca^{2+}, Mg^{2+}) and anions (HCO_3^-, SO_4^{2-}, Cl^-, NO_3^-). These constituents are generally present in all water with concentration from 1 milligram per litre to several grams per litre (seawater). Notice that all these ions have a natural origin, except nitrate.
- nonmetallic minerals, associated with water pollution, including N, P, S (except sulphate) compounds. The nature and concentration of these constituents is highly dependent on the origin of the pollution (urban, agriculture, industries, etc.) and its importance and type (passed, chronic or accidental). The concentrations corresponding are obviously variable, but can reach several hundred milligrams per litre.
- metallic constituents coming from natural origin (ore) or more frequently from anthropogenic pollution. Among them, toxic and heavy metals including hexavalent chromium, cadmium, and mercury must be monitored.

Almost all these ions can be determined with a colorimetric method [1–3].

2. INORGANIC NON METALLIC CONSTITUENTS

N and P compounds are probably the most important inorganic nonmetallic constituents with regard to their environmental effects such as eutrophication of surface water (lakes and rivers). Some of them are directly considered as nutrients (nitrate or phosphate), while others are nutrients precursors (ammonia, organic nitrogen and organic phosphorous). Other constituents must also be considered, such as S compounds because of specific environmental odour problems.

	Ia	IIa		IIIa	IVa	Va	VIa	VIIa	VIII
1	H								He
2	Li	Be		**B** $B_4O_7^{2-}$	**C** CO_3^{2-}, HCO_3^- CN^-	**N** NH_4^+ NO_2^-, NO_3^-	**O** O_2, O_3, H_2O_2, Oxianions	**F** F^-	Ne
3	Na	Mg		Al	Si	**P** HPO_4^{2-}	**S** SO_3^{2-}, HSO_3^- SO_4^{2-}, $S_2O_3^{2-}$, $S_2O_8^{2-}$, SCN^- S^{2-}, HS^-	**Cl** Cl^- ClO^- ClO_3^-	Ar
4	K	Ca	*	Ga	Ge	As	Se	**Br** Br^- BrO_3	Kr
5	Rb	Sr	**	In	Sn	Sb	Te	**I** I^- IO_3^-, IO_4^-	Xe
6	Cs	Ba	***	Tl	Pb	Bi	Po	At	Rn

	IIIb	IVb	Vb	VIb	VIIb	VIIIb	VIIIb	VIIIb	Ib	IIb
* 4	Sc	Ti	V 5	**Cr** CrO^- CrO_4^{2-} $Cr_2O_7^{2-}$	**Mn** Mn^+ MnO_4^-	**Fe** Fe^+ Fe^{2+}	Co	Ni	**Cu** Cu^{2+}	Zn
** 5	Y	Zr	Nb	Mo	Tc	Ru	Rh	Pd	Ag	Cd
*** 6	La	Hf	Ta	W	Re	Os	Ir	Pt	Au	**Hg** Hg^{2+}

FIGURE 1. *Mineral constituents studies in this section (dark box) and in the library (pale box).*

2.1. N compounds

Among all usual parameters for water quality control, nitrogen is probably the most known and monitored, particularly nitrate. N is present in water in reduced form (organic and ammonium nitrogen) and oxidised form (nitrites and nitrates). The evolution of the N compounds depends on physico-chemical and biological mechanisms occurring in natural water or all along treatment processes (Fig. 2).

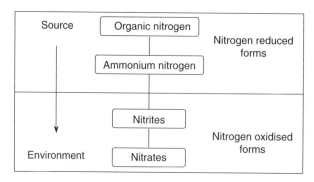

FIGURE 2. *Main nitrogen forms in water.*

The presence of the reduced compounds in surface water is due to natural organic matter and, in major part, to discharges coming from biological wastewater treatment plants for the ammoniacal form. These reduced compounds increase the oxygen demand, resulting in the formation of final stable products as nitrate and are toxic for fish in rivers. Oxidised forms can also be toxic for human beings, through tap water. Nitrate can be chemically reduced in nitrite, which is responsible for the lack of oxygenation of cells, more particularly for babies.

Classical procedures of determination of N compounds are relatively simple for specific forms of nitrogen, as nitrate, nitrite and ammoniacal nitrogen. But most of these methods require an important analytical time (up to 2 h), and often need reagents (for colorimetric or ion chromatographic methods). More particularly for Total Kjeldahl Nitrogen (TKN), including organic and ammoniacal nitrogen, the reference procedure [2] needs minerali-sation and distillation steps and the use of a final titrimetric determination. Moreover, all these methods are sensitive to interferences.

2.1.1. General procedure

A simple methodology, called UV/UV method, has been designed for the quick determi-nation of the different forms of nitrogen in water, nitrite, nitrate, ammonium and TKN [4]. The general procedure is based on several steps (Fig. 3) but includes a UV determination of the oxidised forms and one or two UV photo-oxidation step(s) for the mineralisation of the reduced forms. Neither filtration nor acidification of the sample is needed.

First of all, the oxidised forms of inorganic nitrogen (nitrate, nitrite) are detected by direct UV spectrophotometry (way A). Second, the sample is photo-oxidised with a photo-digester as, for example, the reactor described in Chapter 2 (way B), in the presence of an alkaline oxidant solution (potassium peroxodisulphate buffered to pH 9). Under the influence of UV radiation (in this case, a low-pressure mercury lamp emitting at 254 nm), the global nitrogen is converted into nitrate. The TKN can then be calculated by the difference between the concentration of global nitrogen and the one of oxidised forms previously determined.

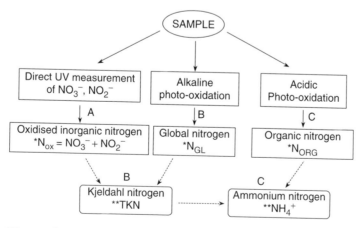

*Measured parameters

**Calculated parameters

FIGURE 3. *General UV/UV procedure for N compounds determination.*

In the third part of the general procedure (way C), the sample is photo-oxidised in the presence of an acidic oxidant solution (pH 2 with sulphuric acid 20%). In these conditions, only organic nitrogen is converted into nitrate. The difference between the concentration of TKN previously calculated and that of organic nitrogen leads to the calculation of ammonium nitrogen concentration.

Thus, all nitrogen forms (N_{ox}, N_{org}, NH_4^+, TKN) can be rapidly and simply determined.

2.1.2. Nitrate measurement

The determination of nitrate is probably the most important application of UV spectropho-tometry for water quality monitoring. The reason is that the spectra of nitrate solutions are very characteristic between 200 and 400 nm in function of the concentration (Fig. 4). Nitrate ion is rather sensitive to UV absorption with a half Gaussian shape for low concen-tration (between 0.5 and 15 mg NO_3^-/L without dilution for 10 mm pathlength). When the concentration increases, an absorption peak appears around 310 nm from 0.2 g NO_3^-/L, without dilution for 10 mm pathlength. In between, no particular shape exists, except the saturation wall for short wavelengths. This typical response is exploited for nitrate determination with different wavelength ranges in function of the expected concentration. Actually, two forms of nitrate ions exist in relatively concentrated solutions (around 5 g/L), with a very slight difference in their UV spectra [5].

Several dozens of works have been published for more than 50 years on the subject, and a few of them are cited hereafter. The proposed methods can be classified according to the exploitation method used, among those presented in Chapter 2:

- The earlier methods are based on simple absorptiometry at one or two wavelength(s) [6–8]. Generally, the use of one absorbance value between 205 and 220 nm, compen-sated by another one between 250 and 275 nm, is proposed. For example, a screening method based on the measurement of absorbance at 220 nm may be used [2], with a correction for dissolved organic matter from a second measurement at 275 nm. For high concentrations (in case of industrial application), a simple absorptiometric method can be envisaged at 310 nm.

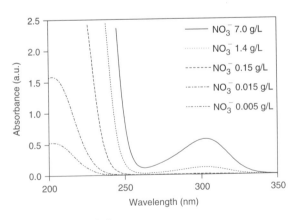

FIGURE 4. *UV spectra of nitrate solutions.*

- The second group states on the interest of the calculation of the second derivative or its estimation with a three-wavelengths measurement, around 220–225 nm [9–11]. These methods are less sensitive to interferences than the first group (supposing that the slope of the UV spectrum of interferences is constant) and can be run on most spectrophotometers.
- The more efficient methods exploit the entire UV spectrum between 200 and 350 nm [12–14], through multiwavelengths procedures. They are based either on PLS algorithm [15] or on the semi-deterministic deconvolution procedure described in Chapter 2. These methods give excellent and very rapid results compared to classical analysis (colorimetry or capillary electrophoresis), but need a specific software for data processing.

Let us insist on one of the main interests of the semi-deterministic deconvolution procedure. The measuring range can vary from 0.5 mg NO_3^-/L to several g NO_3^-/L with the same pathlength (10 mm, for example) without dilution. For the purpose, as the wavelengths window used for the spectrum exploitation has to be adapted to the expected concentration, the precision varies in consequence and can be rather coarse for the intermediate range of concentration. This is the reason why the sample dilution is a good compromise in order to be able to apply the semi-deterministic method, the characteristics of which are presented in Table 1.

For quantitative determination, the semi-deterministic procedure needs to use both a reference spectra basis and the corresponding calibration files (see Chapters 2 and 4). The value of nitrate concentration can then be calculated using the following relation:

$$[NO_3^-] = \alpha_{NO_3}.C_{NO_3} + r$$

where α_{NO_3} is the contribution coefficient value of the reference spectrum corresponding to nitrate, C_{NO_3} the concentration associated with this reference spectrum (expressed if needed in equivalent nitrogen), and r is the quadratic error on the computation of the parameter value.

The procedure has been applied on real samples of water and wastewater, and compared with the values obtained by capillary electrophoresis as reference method (Fig. 5).

The results of the comparison show that the proposed method is well suited for nitrate measurement even in water or wastewater samples with interferences from suspended solids (up to 400 mg$^-$/L) or other dissolved compounds, such as humic substances.

2.1.3. Nitrite measurement

Nitrite ion also presents a specific absorption slightly different from the one of nitrate (Fig. 6). A first peak can be obtained at 210 nm for relatively low concentrations (in fact,

TABLE 1. *Characteristics of the UV method for nitrate determination (without dilution)*

Wavelength range	205–325 nm
Concentration range	0.5–30 mg NO_3^-/L
Detection limit	0.5 mg NO_3^-/L

B. Roig et al.

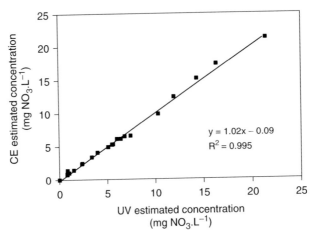

FIGURE 5. Comparison of nitrate analysis by capillary electrophoresis and UV.

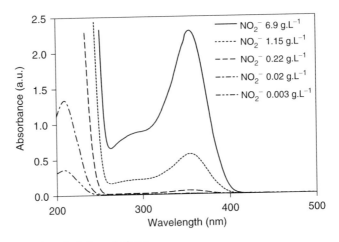

FIGURE 6. UV spectra of nitrite solutions.

rather high with respect to regulation limits), as for nitrate. This first peak appears from concentrations greater than 0.5 mg NO_2^-/L for 10 mm pathlength. A second peak appears when concentration increases at 355 nm from 0.2 g NO_2^-/L, without dilution for 10 mm pathlength, as for nitrate. However, contrary to the nitrate spectrum, a shoulder at 280–300 nm accompanies the second peak. This difference can be used for the analysis of nitrate and nitrite in mixture. Actually, as nitrite is very unstable and is easily oxidised into nitrate, its concentration in water is often very low. Thus, UV spectrophotometry seems to be useful, for example, for industrial applications where nitrite concentration may be higher than 0.5 mg NO_2^-/L. Table 2 presents the main characteristics of the UV method applied for nitrite determination.

TABLE 2. *Characteristics of the UV method for nitrite determination (without dilution)*

Wavelength range	205–325 nm
Concentration range	1–30 mg NO_2^-/L
Detection limit	1.0 mg NO_2^-/L

2.1.4. TKN measurement

Concerning nitrogenous organic compounds, several methods using UV radiation for the mineralisation of samples have been proposed in order to simplify the reference Kjeldhal procedure [16–19]. These methods generally require a classical low-pressure mercury lamp emitting at 254 nm. The use of the far UV radiation of the Hg lamp (185 nm) is made possible by using transparent Suprasil quartz and leads to the improvement of the photodegradation process as compared to the Kjeldahl method (5 to 10 h for Kjeldahl method; 2 to 3 h for direct UV). The use of a strong oxidant with a far UV radiation leads to the drastic decrease of the reaction time (5 to 10 min) [20]. The reaction can be followed by UV spectrophotometry (Fig. 7) with the disappearance of initial compounds and the appearance of final compounds (nitrite and nitrate). The quantification is possible with the semi-deterministic approach by using two sets of reference spectra for the restitution of spectra acquired during the photodegradation. The first basis is the one defined for the estimation of aggregated parameters (see Chapter 4) in water and wastewater. The second basis is simpler as it can include only nitrate and nitrite spectra.

For TKN determination, the photo-oxidation is carried out in the presence of an oxidant solution (potassium peroxodisulphate $K_2S_2O_8$ buffered to pH 9). In this case, organic and inorganic nitrogen compounds are converted into nitrate. The concentration of TKN is equal to the difference between nitrate measured after photo-oxidation ($[NOX]_f$), and nitrate measured before ($[NOX]_i$). It is based on the following relation:

$$[TKN] = [NOX]_f - [NOX]_i$$

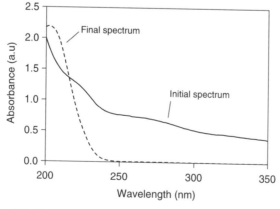

FIGURE 7. *Initial and final spectrum during photodegradation of an urban wastewater sample.*

B. Roig et al.

TABLE 3. *Percentage of conversion into nitrate from N-compounds*

Compounds*	UV method** (percentage recovery)	Compounds*	UV method** (percentage recovery)
Urea	95–100	Glycine	90–95
N-Acetyl-glucosamine	95–100	EDTA	80–90
4-aminophenol	95–100	m-toluidine	80–90
2-nitrophenol	90–100	Atrazine	80–90
4-nitrophenol	90–100	Aniline	70–80
3-aminophenol	90–100	Glutamic acid	70–80

*Concentration between 5 and 50 mg N/L, pH 9.
**5 min irradiation time.

The procedure was first applied on model compounds. Table 3 displays the results of the determination of the nitrogen concentration from various N-containing compounds. The conversion into nitrate from all compounds is quantitative, whatever the concentration, and the conversion times are very short, around 5 min of irradiation time [18].

The conversion into nitrate is possible under UV irradiation without the presence of chemical oxidant, with an increase of the reaction time (around 2 h).

The validation of the procedure has been carried out from the UV/UV estimation of TKN concentration of 80 samples of urban and industrial wastewater. TKN standard method was used as reference method, and the results are shown on Fig. 8. A good adjustment can be observed between the results for a wide concentration range. From few mg N/L to several hundred mg N/L, the comparison between the two methods is good, as well as the precision.

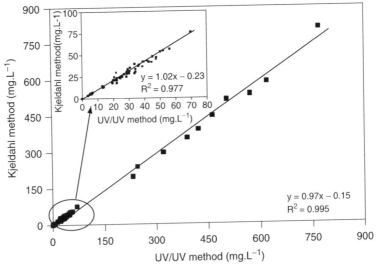

FIGURE 8. *TKN measurement comparison between the Kjeldahl and the UV/UV methods [18].*

TABLE 4. *Characteristics of the UV method for TKN determination (without dilution)*

Wavelength range	205–325 nm
Concentration range	1–20 mg N/L
Detection limit	1.0 mg N/L

Thus, the UV/UV method, the characteristics of which are presented in Table 4, can be used as an alternative way for the reference TKN method.

2.1.5. *Ammonium measurement*

For ammonium determination, two steps of photo-oxidation are needed [21]. The first one allows the measurement of TKN as shown above. Then, a second one is carried out in the presence of an oxidant solution (potassium peroxodisulphate $K_2S_2O_8$), without buffer. In these particular conditions (acidic medium), only organic nitrogen compounds are converted into nitrate, as will be explained afterwards.

The concentration of ammonium can be calculated by the difference between the TKN-estimated concentration ([TKN]) and the one of organic nitrogen ([N_{ORG}]), taking into account an eventual dilution. The relation is the following, where all concentrations must obviously be expressed in the same unit (mg N/L):

$$[NH_4^+] = [TKN] - [N_{ORG}]$$

A total of 80 industrial wastewater samples were analysed with both the standard method (capillary electrophoresis) and the UV/UV one (Fig. 9). A good adjustment can be observed between the results for a wide concentration range (1–100 mg N/L).

Table 5 gives the detection limit and the range of the ammonium measurement.

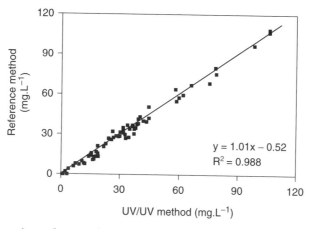

FIGURE 9. *Comparison of ammonium determination for industrial wastewater (reference method is capillary electrophoresis).*

TABLE 5. *Characteristics of the UV method for ammonium determination (without dilution)*

Wavelength range	205–325 nm
Concentration range	1–20 mg N/L
Detection limit	1.2 mg N/L

TABLE 6. *Time analysis (in minutes) for the determination of nitrogen forms*

	NO_x	N_{org}	N_{global}	TKN	NH_4^+
Photo-oxidation	no	10–15	10–15	no	no
Measurement	1	1	1	no	no
Exploitation	1 (dec)	1 (dec)	1 (dec)	1 (cal)	1 (cal)
Total	2	10–15	10–15	1	1

Note: dec: deconvolution; cal: calculation.

The presented UV/UV method allows the simple and reliable determination of all nitrogen forms by using the same technique. Moreover, the total reaction time is very short. A maximum of 40 min is required for all measurements (Table 6).

Some comments can be made concerning the process, and more precisely the impact of the pH on the photodegradation results. Effectively, the photo-oxidation of nitrogenous organic compounds leads to the cleavage of C—N bond resulting in the formation of ammonium radical $NH_2^{+\bullet}$ [22]. A hydrolysis step follows this reaction in order to obtain firstly ammoniacal nitrogen and secondly the final oxidised forms of nitrogen, i.e. nitrite and nitrate.

Moreover, the pH value allows the specific reactivity of organic and inorganic nitrogen. The pH effect can be studied from the comparison of the photo-oxidation of urea (as organic nitrogen) and ammonium chloride (as inorganic nitrogen). It appears that the conversion of urea is weakly modified by the pH and varies between 80 and 95%, whatever the pH value, whereas the conversion yield from the ammonium nitrogen is very weak in acidic medium and becomes quantitative for basic conditions [18].

The presence in wastewater of several organic (solvent) and inorganic (carbonates) compounds may interfere with the photo-oxidation because they are scavengers, inhibiting the action of oxidant radicals. Their action can decrease the conversion yield to 50% in function of the concentration [18].

2.2. P compounds

Phosphorus compounds occur in wastewater under various forms, among which the reduced ones are predominant. The phosphorus compounds of wastewater concern not only some organic forms (natural or anthropic) but also orthophosphate ion and acid hydrolysable phosphate (condensed phosphate) (Fig. 10).

Standard methods [2] used for the determination total phosphorus include several methods. Orthophosphates (PO_4^{3-} and associated forms) are determined by colorimetry, ion chromatography or spectrophotometry (UV-visible). Acid hydrolysis allows the transformation of polyphosphates in orthophosphates, which are measured by one of the mentioned methods. Finally, a chemical digestion, followed by the determination of

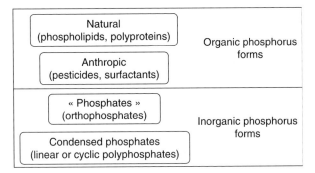

FIGURE 10. *Phosphorous forms in water.*

resulting species (PO_4^{3-}) allows the measurement of organic phosphorus forms. Procedures are time-consuming (more than 5 h) and require strong conditions (acidic medium, high temperature, catalyst).

Some improvements have been proposed, such as the use of photo-oxidation as an alternative to chemical digestion, as it has been successfully used for similar applications [23–26]. In order to minimise the photo-oxidation time, the UV light source (high-, medium- or low-pressure mercury lamp) can be associated with oxidants (hydrogen peroxide or potassium peroxodisulphate). Generally, the converted forms (orthophosphates) are determined by off-line analysis, but can also be measured by flow injection or sequential injection analysis [27,28].

In this section, a simple procedure, based on the use of a UV photo-oxidation module (previously described) and a UV-visible measurement, is presented [4]. Another procedure, based on the use of the alternative vanadomolybdophosphoric acid method with a UV-LED detection at 380 nm can also be envisaged [29].

2.2.1. General procedure

As previously mentioned, phosphorus compounds are commonly classified into orthophosphates (PO_4^{3-}), acid-hydrolysable (condensed) phosphates and organic phosphates. It must be noticed that acid-hydrolysable phosphates (as pyrophosphates) are negligible in sewage [30]. The general procedure illustrated in Fig. 11 includes two main steps: an indirect UV-visible measurement (PO_4^{3-}) and a photo-oxidation step followed by a UV-visible measurement (P_{GL}). First, orthophosphates are determined by spectrophotometric measurement of a phosphomolybdate complex (formed with addition of ammonium molybdate 40 gL^{-1}) using the spectrum deconvolution method.

Then, an oxidant solution (40 g/L potassium peroxodisulphate $K_2S_2O_8$) is added to the sample for the photoconversion (15 min irradiation times) of the organic phosphorus forms into orthophosphates. Orthophosphates are then determined by UV-visible spectrophotometry.

The procedure allows the measurement of orthophosphate (PO_4^{3-}) and global phosphorus (P_{GL}). The determination of the sum of organic and hydrolysable phosphorus ($P_{ORG} + P_{hyd} = P_{GL} - PO_4^{3-}$) is then possible.

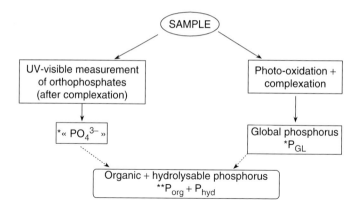

*Measured parameters

**Calculated parameters

FIGURE 11. *Speciation of phosphorus in wastewater.*

2.2.2. Orthophosphates

The measurement of orthophosphates is possible from the phosphomolybdate complex formation. A base of reference spectra has been used with spectra corresponding to molybdate solution (2.4 g/L) and to the phosphomolybdate complex obtained from 10 mg/L of orthophosphate. Notice that in this case the use of a simple multicomponent procedure may be sufficient. The quantification is carried out between 380 and 450 nm. The orthophosphate concentration is given by the product of the phosphomolybdate complex coefficient by the concentration of the corresponding reference spectrum.

Figure 12 shows the comparison between the orthophosphates concentration estimated by deconvolution and the concentration measured by ascorbic acid colorimetry as

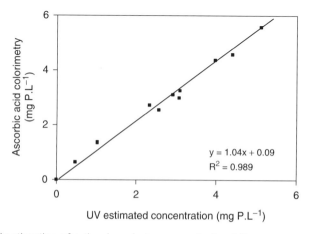

FIGURE 12. *UV estimation of orthophosphate concentration [4].*

TABLE 7. *Characteristics of the UV method for orthophosphates determination (without dilution)*

Wavelength range	380–450 nm
Concentration range	0.05–10 mg P/L
Detection limit	0.05 mg P/L

a reference method. The UV-estimated concentrations are in good agreement with the expected ones and then can be used as an alternative to the standardised method. Even if the characteristics of the method are interesting (Table 7), the detection limit should be lowered to 5 µg/L for trace analysis.

2.2.3. Total phosphorus

The determination of total phosphorus needs a photo-oxidation step in the presence of an oxidant (potassium peroxodisuphate) allowing the transformation of all phosphorus forms into orthophosphates that are complexed (with a molybdate solution) when they are formed [4].

$$\text{Phosphorus forms} \xrightarrow{\text{h}\nu/\text{S}_2\text{O}_8^{2-}/\text{Molybdate}} \text{''PO}_4^{3-}\text{''}$$

Table 8 displays the results for the determination of the orthophosphate concentration from the photo-oxidation of various P-containing compounds. The molybdate solution (1.5 ml) is introduced in the reactor in order to follow the phosphomolybdate complex formation. The conversion yields are quantitative (Table 8) using irradiation time no longer than 15 min.

The UV/UV-visible method was applied for the determination of total phosphorus from raw effluents. The results were compared with those obtained by atomic absorption analysis as the reference method (Fig. 13).

This comparison showed a good correlation ($R^2 = 0.979$). Table 9 presents the characteristic of the described method.

TABLE 8. *Conversion yields obtained from organic phosphorus compounds*

Compounds*	Conversion yields**	Compounds*	Conversion yields**
N (phosphonomethyl) glycine	100	Glucose-1-phosphate	100
Mevinphos	100	AMP	90
Dichlorvos	100	ADP	90
Dibrom	100	ATP	85
Tris (2chloroethyl) phosphate	90		

*Concentration between 1 and 10 mg/L.
**10–15 min irradiation time.

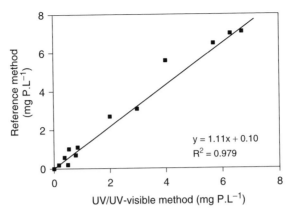

FIGURE 13. *Comparison of total phosphorus measurement by the atomic absorption and UV/UV-visible methods.*

TABLE 9. *Characteristics of the UV method for total phosphorous measurement (without dilution)*

Wavelength range	380–450 nm
Concentration range	0.05–10 mg P/L
Detection limit	0.05 mg P/L

The UV/UV system has been described as an efficient method for the measurement of phosphorus in wastewater. According to the experimental conditions of photo-oxidation or UV measurement, it will be possible to evaluate the major part of phosphorus forms (oxidised, organics, inorganics, hydrolysable) and total phosphorus. Compared to the standardised method, this simple UV/UV-visible method presents some advantages especially in the time consumed, which is six times lower. Moreover, this procedure minimises the consumption of reagents and is realised in softer conditions.

2.3. S compounds

Sulphur occurs in wastewater in various forms. Most of the time, anoxic conditions in urban sewer lead to the production of hydrogen sulphide but, more often, the presence of sulphur compounds is related to industrial discharges, mainly from refineries or petrochemical plants.

Some petroleum contains elemental sulphur, which can occur as hydrogen sulphide (H_2S) and carbonyl sulphide (COS). Sulphur is also present in a wide range of hydrocarbons, largely as mercaptans, organic sulphides and thiophene derivatives [31]. Sulphur compounds tend to be concentrated in the higher boiling fractions of petroleum, are generally corrosive for metals and may poison various catalysts. Stripping water is responsible for the presence of sulphide and mercaptans into crude oil refinery wastewater.

TABLE 10. *Main sulphur species related to water and wastewater (OM: organic matter)*

Sulphur species	Formula	Origin or use
Sulphide ion	S^{2-}	Industrial
Bisulphide ion	HS^-	Industrial/OM reductive degradation
Hydrogen sulphide	H_2S	OM reductive degradation (gas)
Thiosulphate ion	$S_2O_3^{2-}$	Chemical reagent (titration)
Tetrathionate ion	$S_2O_5^{2-}$	Industrial
Sulphate ion	SO_4^{2-}	OM oxidative degradation
Peroxodisulphate ion	$S_2O_8^{2-}$	Chemical reagent (digestion)

Sulphide under dissolved H_2S form is toxic for fish and other aquatic organisms [2] and can be responsible for the decrease of wastewater treatment plants efficiency. Therefore, its concentration needs to be controlled, especially in wastewater from crude oil refineries.

Sulphide can be determined in different media, using various techniques [32–34] that are, unfortunately, often complex or nonrobust (interferences).

Inorganic sulphur compounds are numerous and can be classified, as for nitrogen, between reduced and oxidised forms. Table 10 lists these compounds and their main origin areas, when they are found in water or wastewater.

The reduced forms of sulphur are also called total sulphides, as they are all associated with the acido-basic equilibria of hydrogen sulphide (at 25°C):

$$pKa\ (H_2S/HS^-) = 7.05$$

$$pKa\ (HS^-/S^{2-}) = 12.92$$

Considering the pKa values, the predominant form is the bisulphide ion (HS^-), since the molecular form of hydrogen sulphide (H_2S) is volatile. This observation is true if the pH value of water or wastewater is about 7.

The corresponding UV spectra, the shape of which depends on the pH value, are shown in Fig. 14. An important peak can be noted at 231 nm, related to the presence of hydrogen sulphide and corresponding to the bathochromic shift between the two forms of first acidity (notice that the spectrum of acidic solutions are much less absorbing because of the hypochromic effect and volatility of hydrogen sulphide).

From an analytical point of view, total sulphide includes dissolved H_2S and bisulphide ion HS^-, which are in equilibrium with hydrogen ions. For pH values of wastewater, the S^{2-} form is generally negligible, less than 1% of the dissolved sulphide under pH 10.

Three main methods are commonly used for the determination of sulphide in solution:

- the colorimetric method, with methylene blue, based on the reaction of sulphide, ferric chloride and dimethyl-*p*-phenylenediamine to produce methylene blue, which absorbs at 664 nm [2]
- the iodometric method based on the oxidation of sulphide by iodine in acidic solution followed by a back titration with sodium thiosulphate solution
- the potentiometric method using a selective silver electrode.

Figure 15 shows UV spectra of refinery wastewater containing mineral sulphide (with corresponding DOC around 1000 mg C/L). The value of samples' pH is around 9.

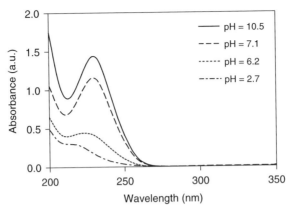

FIGURE 14. *UV spectra of a sulphide solution (HS⁻ 10 mg/L).*

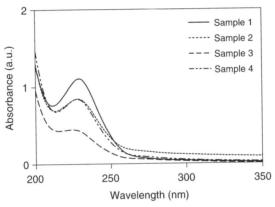

FIGURE 15. *Refinery wastewater UV spectra (dilution 25), with sulphide (respectively, 88, 39, 23, 45 mg S/L).*

The characteristic peak of bisulphide ion appears clearly on the UV spectra (231 nm), despite the matrix sample. The intensity of the UV band is related to the concentration of sulphide.

The use of a UV spectrophotometric procedure can thus be proposed as an alternative method for the determination of inorganic sulphide in water and wastewater. A first method, based on the use of a multiwavelength procedure has been proposed for natural water [35]. The interferences are modelled from an exponential function and the simultaneous determination of total sulphide and iodide is possible. A second method integrates the semi-deterministic deconvolution procedure [36]. The potentiometric method will be chosen as reference for the validation of this last procedure. The UV quantification is carried out by deconvolution (see Chapter 2) between 205 nm and 320 nm. Raw samples were diluted four times to prevent the UV signal saturation for a 10-mm quartz cell. Sulphide (HS⁻) concentration is given by the product of the contribution coefficient of sulphide reference spectrum (replacing nitrate in the previous set) and the corresponding concentration, affected by the dilution factor.

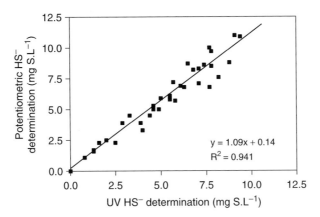

$$y = 1.09x + 0.14$$
$$R^2 = 0.941$$

FIGURE 16. *Validation of sulphide UV determination.*

About 40 wastewater samples from refineries have been used for the validation of the UV method [34]. Figure 16 shows a good linear adjustment between the measured concentration by the potentiometric method and UV determination of sulphide. The characteristics of the UV method are given in Table 11.

Some compounds absorbing close to 231 nm may be present in raw samples and thus may interfere with sulphide spectra. Several compounds, such as, for example, p-chlorophenol (absorption at 227 nm), anionic surfactant, *RBS* commercial product (absorption at 223 nm) and 1-propanthiol (absorption at 239 nm), were tested, and the results show low interference values to the studied compounds [34]. The error of restitution by deconvolution is 5% at maximum with the highest interference associated with anionic surfactant (leading to an error of 4.3%).

The UV method has been described as a simple and reliable procedure for the determination of sulphide in wastewater. Compared with some reference methods, it is less sensitive but do not need any sample preparation (pretreatment, filtration, etc.) and is unaffected by interferences (salinity, suspended matter, organics compounds, etc.).

Finally, the UV determination of sulphate, the knowledge of which is very important for natural water quality, is not yet possible, as this ion in not absorbing, and no simple and robust colorimetric method is available.

2.4. Cl compounds

Chlorine compounds are potentially numerous, as sulphur ones, with several oxidation states. Nevertheless, there is a great difference between the two elements, because

TABLE 11. *Characteristics of the UV method for sulphide determination (without dilution)*

Wavelength range	205–320 nm
Concentration range	0.5 to 15 mg S/L
Detection limit	0.5 mg S/L

all oxidised forms are generally unstable in solution for chlorine compounds. Some of them, e.g. chlorine itself (Cl_2), hypochlorite ion (ClO^-) and chlorine dioxide (ClO_2), are used as oxidant agents for tap water production or swimming pool treatment.

2.4.1. Chloride

Chloride ion is one of the major inorganic anions in water and wastewater because it is the more stable form of chlorine in solution. Chloride concentration is very variable and can be high and not only in seawater. It is used for domestic purpose but rather often for industrial processes. In some applications, the resulting concentration in wastewater can be important and greater than 1 g/L (food and chemical industry, for example).

Figure 17 presents several spectra of industrial water containing chloride as compared to seawater. In all cases, the shape of chloride signal can be observed, characterised by a very high increase of absorbance below 210–220 nm, always convex, in contrast to nitrate. The difference between industrial cooling water UV spectra and seawater is that the first one consists of the presence of some organic matter, responsible for a residual diffused absorption for wavelengths greater than 220 nm.

Several methods can be used for the determination of chloride in water [2]. The argentometric and mercuric nitrate methods are based on the titration of chlorine in the presence of an indicator. Experimental procedures are easy, but many substances may interfere with the results. There are also other methods such as potentiometry, capillary electrophoresis and other automated methods (ferricyanide method or flow injection analysis).

UV spectrophotometry can be proposed for the quick estimation of high chloride concentration (above 500 mg/L).

The semi-deterministic method used for UV spectra exploitation method can be used by integrating the spectrum of a chloride solution on the basis of reference spectra. The concentration is then calculated by multiplying the corresponding chloride concentration with the contribution coefficient of the chloride spectrum.

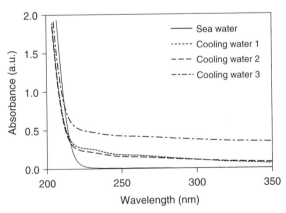

FIGURE 17. *UV spectra of cooling and seawater (dilution 5). Chloride concentrations are between 15 and 22 g/L.*

TABLE 12. *Characteristics of the UV method for chloride estimation (without dilution)*

Wavelength range	200–320 nm
Concentration range	0.5–6 g/L
Detection limit	0.5 g/L

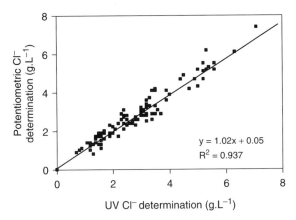

FIGURE 18. *Comparison of UV estimation of chloride in industrial water with the potentiometric method.*

The characteristics of the method are given in Table 12. A comparison between this UV procedure and the reference potentiometric method has been carried out on 110 industrial samples. Figure 18 shows the quality of the adjustment between the two sets of results.

2.4.2. Hypochlorite

Chlorine may also be found in water under the hypochlorite ion, which is a strong oxidant used for water and housing disinfection. It is used for tap water production as well as for swimming pool water treatment or, in some cases, for the oxidation of odorous compounds. This product (commonly found as "Eau de Javel") also has a persistent disinfecting action as long as hypochlorite is in the solution. The residual concentration must thus be sufficient but not in large excess, because of the formation of organohalogeno compounds such as chloramines, which are responsible for eye irritations and can even be toxic.

The presence of hypochlorite ion can easily be monitored by UV spectrophotometry for relatively high concentrations, because its UV spectrum presents, contrary to the one of chloride ion, an important Gaussian-like peak centred on 290 nm. Applications for oxidation process control using hypochlorite, gaseous chlorine or chlorine dioxide are possible using UV spectrophotometry. The spectrum exploitation is possible with the semi-deterministic method (by including the corresponding reference spectrum, as previously) but a simpler absorptiometric procedure can be used at 290 nm, if interferences (organic compounds, suspended solids, etc.) are negligible.

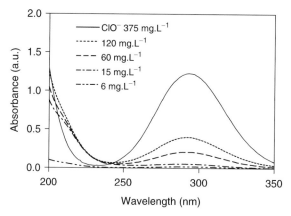

FIGURE 19. *UV spectra of water (swimming pool and deodorisation) containing hypochlorite.*

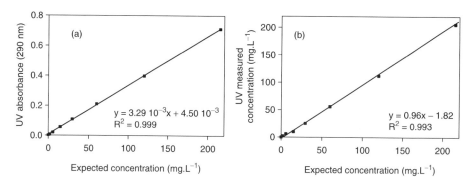

FIGURE 20. *Hypochlorite detection by absorptiometry (a) or by semi-deterministic method (b).*

Figure 19 shows some examples of swimming pool and deodorisation water. A comparison has been carried out between the concentration obtained by UV spectrophotometry, either simple absorptiometry at 290 nm or with the semi-deterministic approach. Figure 20 shows the right adjustment for the results for ten samples of real water with standard additions of hypochlorite.

Table 13 presents the characteristics of the UV methods for a pathlength of 10 mm. Using a 100-mm pathlength lead to a detection limit of 0.5 mg/L, a value acceptable for some regulation applications.

3. METALLIC CONSTITUENTS

The constant evolution of water quality standards implies the development of faster and cheaper analytical procedures. These are more particularly adapted for online measurement, for the frequent analysis of metallic constituents or for the estimation of

TABLE 13. *Characteristics of the UV methods for hypochlorite measurement*

Wavelength range	220–325 or 290* nm
Concentration range (without sample dilution)	0–600 mg/L
Detection limit	5 mg/L

*Usable in case of no interference.

global parameters. The cheapest and simplest methods for the determination of metallic constituents are colorimetric procedures, which are less precise and selective than instrumental reference methods as GF-AAS or ICP-AES. Other instrumental techniques based on electrochemical principles are also proposed for water quality monitoring, but they are relatively expensive, sensitive to interferences and, thus, inappropriate for wastewater monitoring.

The improvement of UV-visible spectrophotometers has led to a renewal in the colorimetric procedures, actually absorptiometric, with the possibility of simultaneous metallic compounds determination. Based on the reaction between a ligand and some metallic elements, these complexometric procedures use absorptiometry for final detection. The characteristics of these methods are also improved with regard to classical colorimetric ones, but they often need some pretreatment steps for preconcentration or interferences removal.

Before considering some of these complexometric procedures with final UV-visible detection, designed for the simultaneous determination of metallic constituents, a more simple approach must be considered. If some metallic constituents absorb in the UV-visible region, it is interesting to check the feasibility of a direct measurement without reagent. This is studied in the first part with the determination of some chromium ions.

3.1. Chromium (direct measurement)

3.1.1. Hexavalent chromium

Chromium VI is one of the major contaminants in industrial wastewater, particularly for metal processing activities, such as electroplating, for example. Because of its toxicity, its maximum concentration in treated wastewater must generally be 50 μg/L before being discharged into a receiving medium.

There exist several analytical methods for chromium VI determination, often very complex [37]. One of the most simple and widespread is the diphenylcarbazide colorimetric method. Its detection limit is about 1 μg/L, but the procedure is time consuming.

Another type of method is based on the spectrophotometric properties of Cr (VI), sometimes used as standard solution [38]. Moreover, the spectrophotometric determination of Cr (VI) is well known in the field of water examination, as it can be used for COD alternative measurement [39] (see Chapter 4). Two other UV-visible spectrophotometric methods have been proposed. The first one [40], designed for natural water, uses the peak height measurement at 372 nm, for a basified sample (pH > 9). The peak height is calculated from the absorbance values at three wavelengths (310, 372 and 480 nm), taking into account a very simple third-degree-polynomial interference signal.

The second UV-visible spectrophotometric method [41] was proposed for natural and urban wastewaters with a more general mathematical compensation of interferences. Both UV-visible spectrophotometric methods are rapid and simple with a detection limit of around 5 μg/L for a 50-mm optical pathlength. Unfortunately, they cannot be applied for industrial wastewater survey, as the presence of specific compounds cannot be modelled by the mathematical tools used for interferences removal.

The application of the semi-deterministic approach, described in Chapter 2 and applied before in this chapter, can be used for the determination of hexavalent chromium. Exceptionally, the optical pathlength is 50 mm for this application because of the regulation compliance constraint. Contrary to the previous determination (nitrate, sulphide, hypochlorite, etc.), this application must take into account the pKa value (about 5.8) of the dichromate in equilibrium with chromate:

$$Cr_2O_7^{2-} + 3H_2O \rightleftarrows 2CrO_4^{2-} + 2H_3O^+$$

As the pH of industrial wastewater must be around the pKa value (before neutralisation), both chromate and dichromate spectra must be chosen on the basis of reference spectra, taking into account the influence of pH on chromium VI speciation (Fig. 21). The two first reference spectra correspond to chromium VI solutions of pH 3.5 and 9.5, respectively.

Figure 22 shows real spectra from industrial, wastewater and natural sample containing hexavalent chromium at different concentrations.

The spectra of Fig. 21 show two absorption peaks depending on pH, in the UV region and a slight residual absorbance in the visible one. Bathochromic and hyperchromic effects can be noticed between the acidic and basic media. The resulting visible colour in the sample containing hexavalent chromium for neutral pH is orange.

A comparison between the diphenylcarbazide and UV-visible method for Cr (VI) determination has been carried out for more than 50 samples of different origins (44 industrial wastewaters, 8 urban wastewaters and 4 natural waters), the concentration of which

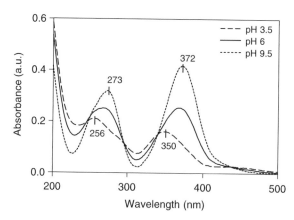

FIGURE 21. *Absorption spectra of a $K_2Cr_2O_7$ solution (1 mg/L of Cr, pathlength 50 mm) for different pH.*

varies between 0 and almost 1 mg/L. Figure 22 shows some examples of the sample spectra. The results of the comparison are shown in Fig. 23, and the adjustment is quite good.

Table 14 presents the characteristics of the method.

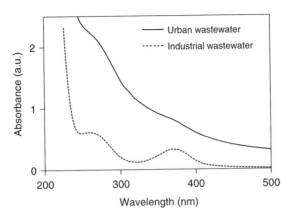

FIGURE 22. *UV-visible spectra of wastewater containing hexavalent chromium.*

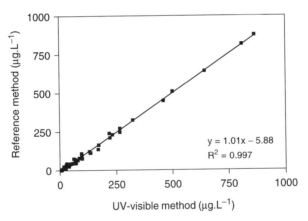

FIGURE 23. *Comparison of the diphenylcarbazide and UV-visible method for Cr (VI) determination, for water and wastewater samples (pathlength 50 mm).*

TABLE 14. *Characteristics of the UV-visible method for Cr (VI) determination (pathlength 50 mm)*

Wavelength range	300–450 nm
Concentration range	5–1000 μg/L
Detection limit	5 μg/L

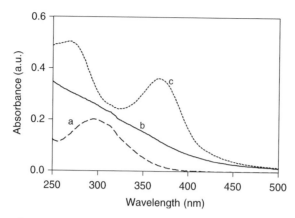

FIGURE 24. *Absorption spectra (pathlength 50 mm) of synthetic solutions containing 1 mg/L of iron (a-pH 3, b-pH 9, c-pH 9 and addition of Cr (VI)).*

As the method is designed for industrial applications, the interference of some metallic compounds must be checked. Indeed, different ions, such as copper (II), iron (III), lead (II) and mercury (II), for example, potentially existing in electroplating wastewater, may interfere with the determination of Cr (VI). Except in the presence of Fe^{3+}, the measurement of Cr (VI) gives an error generally lower than 3% for 0.5 or 1 mg/L of metallic ion, and for greater concentrations (up to 10 mg/L), the error is negligible for Cu^{2+}, and lower than 15% for the other ions [42].

The reason why Fe^{3+} interferes is shown in Fig. 24. Even for a concentration of 1 mg/L, Fe^{3+}, which is the most probable form of dissolved iron in water, absorbs around 300 nm in acidic medium. In this case, the quadratic error of the deconvolution is too high. An increase of the pH value up to 9 (with some drops of NaOH 1M, for example), leads to the precipitation of the hydroxide form, the spectrum of which is very close to the reference of suspended solids. With this simple pretreatment, the error in Cr (VI) determination between the diphenylcarbazide and the UV-visible methods becomes lower than 2.5%.

3.1.2. Trivalent chromium

Trivalent chromium (Cr^{3+} ion) is not as toxic as the hexavalent form and can be directly detected by visible spectrophotometry, its UV absorption being nonspecific. Figure 25 presents the UV-visible spectra of the trivalent form and shows the great difference with one of the hexavalent form. Two peaks of absorbance are noted at 433 and 600 nm, giving a green colour to the solution.

The main application of the optical properties of Cr^{3+} ion is the final colorimetric determination of COD with the measurement of absorbance at 600 nm, maximum of absorbance in acidic medium (see Chapter 4).

As a consequence, the direct spectrophotometric determination of Cr^{3+} ion in water and wastewater is limited to highly concentrated samples.

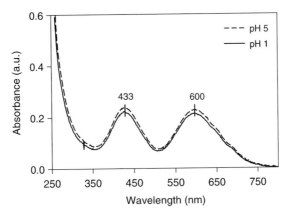

FIGURE 25. *UV-visible spectrum of trivalent chromium (after filtration of saturated solution).*

3.2. Metallic constituents determination by complexometry

Several procedures for the simultaneous determination of metallic constituents have been proposed (Table 15). Based on the complexometric reaction between a ligand and some metallic elements, the resulting spectrum is exploited in order to determine the corresponding concentrations. The exploitation methods are classical, using derivative or multicomponent spectrophotometry. The choice of the different reagents (complexant, buffer, etc.) is obviously dependant on the metallic ion to be determined, but also on

TABLE 15. *Recent complexometric procedures for the simultaneous determination of metallic constituents (examples)*

Metallic constituent	Complexant	Exploitation method	Ref.
Co, Cu, Pb, Mn, Ni, Zn, Fe	PAR	First-, second- and third-derivative spectrophotometry	[43]
Ni, Co, Cu, Fe	PAR	Derivative spectrophotometry	[44]
Cd, Co, Mn, Ni, Zn, Pb	PAR	Multiple linear regression	[45]
Ca, Mg	PAR	Multicomponent analysis	[46]
Cr, Cu	MEDTA	Derivative spectrophotometry	[47]
Cu, Co	MEDTA	Derivative spectrophotometry	[48]
Fe, Rh	Diphenyl phenantholin and glycol	Second-derivative spectrophotometry	[49]
Al, Fe	Hematoxylin + cetyltrimethylammonium bromide	First- and second-derivative spectrophotometry	[50]
Fe, Ag, Mn	Rhodamin B	Measurement at 555 nm	[51]
Cu, Fe	BBT, bathocuproin	Second-derivative spectrophotometry	[52]
Cu, Pd	Oxazolin	Spectrum and first-order derivative spectrophotometry	[53]

their optical properties (principally in the visible region). The most employed complexants are PAR (4-pyridyl-2-azo resorcinol), presented in Chapter 4, and EDTA and its derivatives. Other complexants are used as can be seen in Table 15.

Concerning the choice of the metallic constituents, the majority of studies deal with the lighter constituents (located in the first four lines of the periodic table). This is due more to the reactivity of constituents with a given complexant than to an environmental interest (heavy metals).

Here we present another more precise complexometric method for the simultaneous determination of Cu, Fe, Hg in water and wastewater [54,55], rather different from the previous ones. On one hand, the complexant solution is based on the use of dithizone (Diphenyl-1,5-thiocarbazone), which was widely used for metal analysis up to the seventies [56,57]. On the other hand, the final determination is coupled with the spectral semi-deterministic method and is applied on large part of the UV-visible spectrum.

This main reagent is a mixture of dithizone as principal ligand (i.e. chelating agent of the studied elements), EDTA employed as masking agent, BHA (butylhydroxyanisol), glycine, ethanol and distilled water. This reagent presents potential applications in natural waters and effluents.

Taking into account the possible metallic hydroxides precipitation and the potential presence of natural complexes (with humic substances in natural waters, or with specific compounds in industrial wastewater) [58], the working pH is fixed at about 2.8, with a glycine solution for the stabilisation of the dithizone spectrum. Notice that the glycine and its eventual metallic complexes are almost transparent in the UV region, and do not change the general shape of the spectra of the final mixture [59]. Similarly, EDTA (in all the wavelength range) and BHA (transparent above 310 nm), alone or complexed, do not interfere with the UV signal except for copper and iron complexes, responsible for a great absorbance in the UV region.

The quantification of the metallic constituents studied (Cu (II), Hg (II) and Fe (II)) is carried out by the spectral semi-deterministic method previously described or by any multicomponent procedure. The choice of the wavelength range depends on the studied element and on the corresponding basis of reference spectra. Moreover, the optical pathlength is either 10 or 50 mm for high or low concentration, respectively, with two reactive solution concentrations [58].

The UV-visible spectra of the metallic complexes constitute the basis of reference spectra used for the calculation of Cu (II), Hg (II) and Fe (II) concentrations. Figure 26 shows these reference spectra (completed by a blank of the reagent), and shows that the UV region is of poor interest for the purpose, taking into account the optical properties of the chosen reagents. Each of these spectra corresponds to a given concentration of a metallic constituent, which serve for the final concentration calculation of a sample. This value is dependant on several parameters, among which is the concentration of the studied metal, related to the stoichiometric equilibrium (with dithizone) and the dilution rate of the sample in the sample–reagent mixture. The metallic constituent concentration of an unknown sample corresponds to the product of this value by the contribution coefficient related to the corresponding dithizonate spectrum in the linear combination.

For natural waters and/or effluent applications, some interference may occur with the presence of natural or anthropogenic chelating agents. For example, humic substances can compete with dithizone for metallic constituent complexation in natural waters [60]. In this case, the degradation of organometallic complexes must be effective before the analytical determination. A photodegradation step, with a simple device as the one already proposed in this chapter for N and P compounds determination, can be used

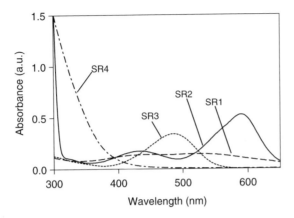

FIGURE 26. *Reference spectra basis for the computation of Cu (II), Hg (II) and Fe (II) concentrations for the high-concentration mixture (10 mm cell pathlength) (SR1: Cu (II)-dithizonate, SR2: blank of the reagent, SR3: Hg (II)-dithizonate, SR4: Fe (II)-EDTA complex) [58].*

for the purpose. However, contrary to the previous procedure using an oxidative reagent, the decomplexation can be carried out without the reagent [58].

The characteristics of the method for high- and low-concentration mixtures are presented in Table 16. The precision values are calculated for the middle of their respective range, and the sensitivities correspond to the smallest concentration difference data that can be obtained, depending both on the spectrophotometer resolution (10^{-3} a.u.) and on the precision for the calculation of the contribution coefficient values. The detection limit values are three times the standard deviation of the blank.

Several experiments have been carried out on samples of natural waters and industrial wastewaters (12 for Cu, 9 for Hg and 14 for Fe), with standard addition for some samples [55]. The analysis of the samples was performed both by a reference method (graphite furnace AAS for copper and iron, cold vapour AAS for Hg), and by the proposed spectrophotometric method.

Figure 27 shows the comparison results between the two methods for each metallic constituent. The correlation coefficients of the related regression lines are satisfactory and greater than 0.95 for the three studied metallic constituents, and the intercept values are very small compared to the concentration ranges.

TABLE 16. *Characteristics of the proposed method*

	Wavelength (nm)	Working range ($\mu g.L^{-1}$)	Detection limit ($\mu g.L^{-1}$)	Precision (%)
Copper	300–650	0–30	3	3.0
		0–600	100	5.6
Mercury	300–650	0–100	2	3.2
		0–5000	80	5.0
Iron	300–650	0–1500	45	0.5
		0–12,500	1200	0.4

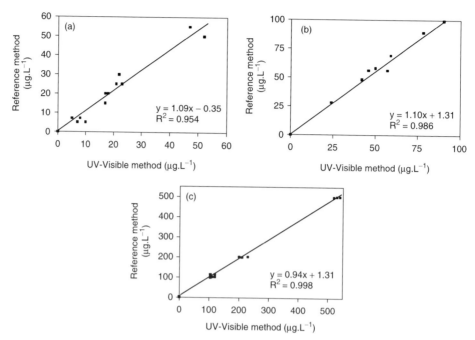

F<small>IGURE</small> 27. *Comparison between the UV-visible spectrophotometric method and the corresponding reference methods for (a) copper, (b) mercury and (c) iron determination, with or without standard additions of metals in sample.*

With respect to normalisation or regulation constraints, the high-concentration procedure (10 mm pathlength and corresponding reagent) is suitable for industrial wastewater control, the low concentration method being more adapted for natural water samples since metallic constituent concentrations are usually very low.

References

1. J. Mendham, R.C. Denney, J.D. Barnes, M.J.K. Thomas, *Vogel's Quantitative Chemical Analysis*, 6th edition, Prentice Hall, New York (1999).
2. A.E. Grennberg, L.S. Clesceri, A.D. Eaton, *Standards Methods for the Examination of Water and Wastewater*, A.P.H.A, 18th edition, Baltimore (1992).
3. J. Rodier, *l'Analyse de l'Eau, Eaux Naturelles, Eaux Résiduaires*, 8th edition, Dunod Eds., eau de mer, Paris (1996).
4. B. Roig, C. Gonzalez, O. Thomas, *Talanta*, 50 (1999) 751.
5. V. Simeon, V. Butorac, V. Tomisic, N. Kallay, *Phys. Chem. Chem. Phys.*, 5 (2003) 2015.
6. R.C. Hoather, *Wat. Treat. Exam.*, 2 (1952) 9.
7. J.P. Rennie, A.M. Summer, F.B. Basketter, *Analyst*, 104 (1979) 837.
8. K.C. Thompson, M. Blankley, *Analyst*, 109 (1984) 1053.
9. J. Simal, M.A. Lage, I. Iglesias, *Analyst*, 68 (1985) 962.
10. M.A. Ferree, R. Shanon, *Water Res.*, 35 (2001) 327.
11. I.E. Kalinichenko, L.N. Demustskaya, *J. Anal. Chem.*, 59 (2004) 240.

12. O. Thomas, S. Gallot, N. Mazas, Fres. *J. Anal. Chem.*, 338 (1990) 238.
13. O. Thomas, F. Theraulaz, M. Domeizel, C. Massiani, *Environ. Technol.*, 14 (1993) 1187.
14. M. Karlsson, B. Karlberg, R.J.O. Olsson, *Anal. Chim. Acta*, 312 (1995) 107.
15. G. Langergraber, N. Fleishmann, F. Hofstadter, *Water Sci. Technol.*, 47 (2003) 63.
16. F.A. Armstrong, P.M. William, J.D.H. Stricland, *Nature*, 21 (1966) 481.
17. B.A. Manny, M.C. Miller, R.G. Wetzel, *Limnol. Oceanogr.*, 16 (1971) 71.
18. J.H. Lowry, K.H. Mancy, *Water Res.*, 12 (1978) 471.
19. H. Kroon, *Anal. Chim. Acta*, 276 (1993) 287.
20. B. Roig, C. Gonzalez, O. Thomas, *Anal. Chim. Acta*, 389 (1999) 267.
21. B. Roig, F. Pouly, C. Gonzalez, O. Thomas, *Anal. Chim. Acta*, 437 (2001) 145.
22. G.K.C. Low, S.R. Mc Evoy, R.W. Matthews, *Environ. Sci. Technol.*, 25 (1991) 460.
23. F.A.J. Amstrong, S. Tibbitts, *J. Mar. Biol. Ass.* UK., 48 (1968) 143.
24. A. Henriksen, *Analyst*, 95 (1970) 601.
25. L.T.H. Goosen, J.G. Kloosterboer, *Anal. Chem.*, 50 (6) (1978) 707.
26. A.N. Shkil, A.V. Krasnushkin, I.T. Gavrilov, *J. Anal. Chem.*, USSR, 45 (8) (1990) 1165.
27. I.D. McKelvie, B.T. Hart, *Analyst*, 114 (1989) 1459.
28. O. Thomas, F. Theraulaz, V. Cerda, D. Constant, P. Quevauviller, *Trends Anal. Chem.*, 16 (1997) 419.
29. M. Bowden, M. Sequiera, J.P. Krog, P. Gravesen, D. Diamond, *J. Environ. Monit.*, 4 (2002) 767.
30. D. Jolley, W. Maher, P. Cullen, *Water Res.*, 32 (3) (1998) 711.
31. Des W. Connell, *Basic Concepts of Environmental Chemistry*, Lewis Publishers, New York (1997).
32. R. Al Farawati, C.M.G. Van Den Berg, *Marine Chem.*, 57 (1997) 277.
33. J.L. Wilcox, R. Del Delumyea, *Anal. Lett.*, 27 (1994) 2805.
34. A.G. Howard, C.Y. Yeh, *Anal. Chem.*, 70 (1998) 4868.
35. E.A. Guenther, K.S. Johnson, K.H. Coale, *Anal. Chem.*, 73 (2001) 3481.
36. F. Pouly, E. Touraud, J.-F. Buisson, O. Thomas, *Talanta*, 50 (1999) 737.
37. V.M. Rao, M.N. Sastri, *J. Sci. Ind. Res.*, 41 (1982) 607.
38. C. Burgess, A. Knowles, *Standard in Absorption Spectrometry*, Chapman and Hall, Londres (1981).
39. O. Thomas, N. Mazas, *Analusis*, 14 (1986) 300.
40. A. Oumedjbeur, O. Thomas, *Analusis*, 17 (1989) 221.
41. O. Thomas, S. Gallot, E. Naffrechoux, *Fres. J. Anal. Chem.*, 338 (1990) 241.
42. E. Bobrowska-Grzesik, A.M. Grossman, *Fres. J. Anal. Chem.*, 354 (1996) 498.
43. L.L. Kolomiets, L.A. Pilipenko, I.M. Zhmud, I.P. Panfilova, *J. Anal. Chem.*, 54 (1999) 28.
44. E. Gomez, J.M. Estela, V. Cerda, M. Blanco, *Fres. J. Anal. Chem.*, 342 (1992) 318.
45. E. Gomez, J.M. Estela, V. Cerda, *Anal. Chim. Acta*, 249 (1991) 513.
46. A. Cladera, E. Gomez, J.M. Estela, V. Cerda, *Intern. J. Environ. Anal. Chem.*, 45 (1991) 143.
47. H. Seco-Lago, J. Perez-Iglesias, J.M. Fernandez-Solis, J.M. Castro-Romero, V. Gonzalez-Rodriguez, *Fres. J. Anal. Chem.*, 357 (1997) 464.
48. J.M. Castro-Romero, J.M. Fernandez-Solis, M.H. Bollain-Rodriguez, F. Bermejo-Martinez, *Microchem. J.*, 43 (1991) 104.
49. M. Toral, I.P. Richter, A.E. Tapia, J. Hernandez, *Talanta*, 50 (1999) 183.
50. Y.A. El-Sayed, Fres. J. Anal. Chem., 355 (1996) 29.
51. Y.Z. Ye, H.-Y. Mao, Y.-H. Chen, *Talanta*, 45 (1998) 1123.
52. M.I. Toral, P. Richter, C. Rodriguez, *Talanta*, 45 (1997) 147.
53. A.A.Y. El-Sayed, M.A.A. Rahem, A.A. Omran, *Anal. Sci.*, 14 (1998) 577.
54. F. Theraulaz, O. Thomas, *Mikrochim. Acta*, 113 (1994) 53.
55. B. Coulomb, F. Theraulaz, V. Cerda, O. Thomas, *Quim. Anal.*, 18 (1999) 255.
56. E.B. Sandell, H. Onishi, *Photometric Determination of Traces of Metals: General Aspect.* 4th edition, John Wiley and Sons, New York (1978).
57. B.W. Budesinsky, M. Sagat, *Talanta*, 20 (1973) 228.

58. J.W. Moore, S. Ramamoorthy, *Heavy Metals in Natural Waters: Applied Monitoring and Impact Assessment*, Springer-Verlag, New York (1984).
59. F. Theraulaz, PhD Thesis, University of Savoie, France (1993).
60. J. Buffle, *Complexation Reactions in Aquatic Systems: An Analytical Approach*, Ellis Horwood Limited, Chichester (1988).

CHAPTER 6

Physical and Aggregate Properties

M.-F. Pouet[a], N. Azema[b], E. Touraud[c], O. Thomas[a]

[a]Observatoire de l'Environnement et du Développeme nt Durable, Université de Sherbrooke, Sherbrooke, Québec, J1K 2R1, Canada; [b]Centre des Matériaux de Grande Diffusion, Ecole des Mines d'Alès, 6 Avenue de Clavières, 30319 Alès Cedex, France; [c]Laboratoire Génie de l'Environnement Industriel, Ecole des Mines d'Alès, 6 Avenue de Clavières, 30319 Alès Cedex, France

1. INTRODUCTION

Parameters previously described in Chapters 3 to 5 are related to either specific compounds or groups of compounds, always in solution. Other parameters must be considered to complete the physico-chemical characterisation of water and wastewater. These parameters can be quantified either by electrical sensors (or similar) such as temperature, redox potential or conductivity (which is actually related to the sum of conducting species), or by optical methods for the determination of colour, turbidity or suspended solids. The latter parameters using UV-visible light are presented in this chapter.

The pollution load of water and wastewater is often associated with the presence of floating, coarse and particulate matters. This type of pollution is particularly important because of its consequences in terms of deposition, clogging and anaerobic degradation, as well as its adsorption potential for metallic or organic compounds, or microorganisms. In fact, due to the loss or dispersion of floating and coarse material during transportation, the particulate fraction, including suspended solids, remain the main parameters for the survey. Indeed, colloidal and particulate matters in wastewater have a great influence on the performance of wastewater treatment plants: each operation unit, such as settling, biological or chemical treatment, is affected by the phenomena of agglomeration or dispersion of colloids [1]. Therefore, an understanding of these different phenomena and the development of methods allowing the characterisation and quantification of solids in sewage are necessary for the optimisation of treatment processes and for the evaluation of their performances [2].

Wastewater is often a mixture of organic and mineral pollutants [3]. For example, urban sewage contains both anthropogenic and natural contaminants whose size distribution is very wide (Fig. 1).

Four families of compounds are usually defined to describe the pollution fractions contained in urban wastewater: soluble fraction (<0.001 μm), colloidal fraction (0.001–1 μm), supracolloidal fraction (1–100 μm) and settleable fraction (>100 μm) [4].

The measurement of total suspended solids (TSS) includes supracolloids and settleable matter. Before considering the TSS measurement, the characterisation of which

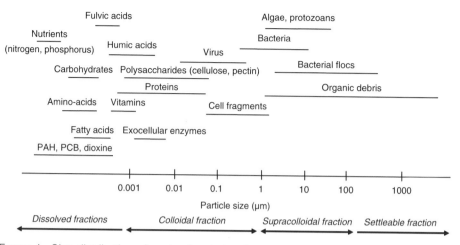

FIGURE 1. *Size distribution of contaminants in urban wastewater (adapted from [6]).*

is very important with respect to their environmental impact, the problem of the border between dissolved and solid phases has to be discussed. Depending on the standard method used for TSS measurements, the result can vary with the cut off size of the filtration systems [5]. Moreover, a part of the colloidal fraction can be included. The colloidal state can thus be defined as an intermediate phase between the solid state (i.e. material of a size greater than 1 μm or 1.2 μm) and the dissolved phase (i.e. compounds smaller than 0.001 μm). This definition is strongly linked to the character-istics of the usual separation devices. Besides, it is possible to distinguish between the supracolloidal fraction (nonsettleable matter of a size larger than 1 μm) and smaller colloids.

Finally, UV spectrophotometry integrates different combined optical phenomena such as physical absorption (particles absorption, diffusion, refraction, diffraction) and chemical absorption. This is the reason why optical responses of wastewater are complex and difficult to interpret.

2. COLOUR

2.1. Determination of colour

Colour is related to the presence of dissolved compounds including chromophores in their structure. Table 1 presents the wavelength regions that correspond to the colours of the visible spectrum.

TABLE 1. *Colours and their complements*

Wavelength (nm)	430	480	540	580	620	650
Colour	Violet	Blue	Green	Yellow	Orange	Red
Complement	Yellow	Orange	Red	Violet	Blue	Green

TABLE 2. *Some methods for the spectrophotometric determination of colour*

Method	Principle	Result	Comments
USEPA 1	Calculation of tristimulus values from transmittance values	Dominant wavelength (nm) Hue, luminance (%) Purity (%)	Use of a set of wavelengths (between 410 and 670 nm)
USEPA 2	Tristimulus filter method	Dominant wavelength (nm) Hue Luminance (%) Purity (%)	Use of three special tristimulus light filters (corresponding to the following wavelengths: 590, 540 and 438 nm)
ISO 7887	Absorbance measurement	Coefficient of spectrum absorption	Use of three wavelengths: 436, 525 and 620 nm

A colour seen by the human eye is in fact the complementary colour, i.e. resulting from the polychromatic light absorption of a solution for a given wavelength [7] (cf. Chapter 1). Coloured substances show a strong absorption in the visible region ($\varepsilon > 10^3$ mole/L/cm). The lower sensitivity limit of the human eye is approximately 380 nm, bordering between the UV and visible regions. Colour is defined by hue (the name of the colour), luminance (relative lightness or darkness of the colour) and saturation (purity of the colour) [6].

Colour determination can be carried out with several reference methods, generally based on the sample optical properties in the visible region (Table 2). The examination of the different procedures leads to the conclusion that, except for the USEPA 1 method, which uses several sets of three wavelengths, the others are limited to the choice of the wavelengths to be considered and give less useful results, which are apparently not very close to the significance of the parameter. Almost all methods can be automatically performed by a PC-controlled spectrophotometer, provided the bandwidth of the instrument is adapted for the measurement.

2.2. Relation between colour and visible absorbance

Figure 2 shows the UV spectrum of an industrial chemical product used for the diagnosis of metallic surface defaults. In case of a small crack, for example, the default remains coloured in red. The UV-visible spectrum shows an important peak at 552 nm, probably due to the presence of an unknown azo-dye.

The origin of the colour can be related to one coloured product or pollutant, or to natural constituents (humic substances, metallic ions or complexes, for instance). The colour is well defined when only one coloured solute is present; but, generally, several compounds are responsible for a mixture of colours giving a broad visible spectrum (Fig. 3).

Tables 3 and 4 present the tristimulus coordinates and the corresponding absorption coefficients of the different samples presented in Figs. 2 and 3. Some observations can be drawn. Except for the industrial product, the corresponding colour characteristics are rather close, even if the visible spectra show some differences. This is particularly true for the tristimulus method which gives an integrated response more adapted for dyes than for water or wastewater. The ISO method seems to discriminate more efficiently the colours but needs a more complete comparative study for each absorption coefficient.

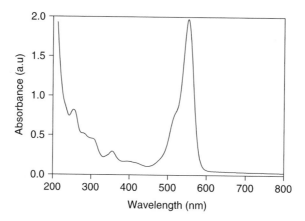

FIGURE 2. *UV-visible spectrum of an industrial product used for metallic surface checking.*

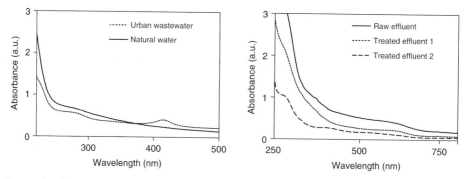

FIGURE 3. *Spectra of coloured solutions and waters.*

TABLE 3. *Results of samples of Figs. 1 and 2 (according to USEPA method)*

Sample	Tristimulus coordinates		Dom. λ	Purity	Colour hue
	X	Y			
Industrial product[a]	0.341	0.196	540	55	Reddish-purple
Natural water[b]	0.329	0.335	580	<10	Yellowish-orange
Urban wastewater[b]	0.354	0.375	574	28	Greenish
Textile wastewater[b]	0.352	0.350	580	21	Yellowish-orange
Treated wastewater 1[b]	0.332	0.340	577	12	Yellow
Treated wastewater 2[b]	0.329	0.331	581	<10	Yellowish-orange

[a]Figure 2.
[b]Figure 3.

TABLE 4. *Spectra absorption coefficients. Results of samples of Figs. 1 and 2 according to ISO 7887*

Sample	α (436)	α (525)	α (620)
Industrial product[a]	12.1	84.4	4.8
Natural water[b]	20.5	13.0	8.2
Urban wastewater[b]	29.8	4.8	3.0
Textile wastewater[b]	64.4	49.2	38.8
Treated wastewater 1[b]	35.4	24.8	21.1
Treated wastewater 2[b]	23.5	16.9	11.1

[a]Figure 2.
[b]Figure 3.

3. PHYSICAL DIFFUSE ABSORPTION

3.1. Some elements on diffusion of light by particles

When a particle is illuminated by a beam of light, it reflects light in all directions. This is the diffusion phenomenon, involving three mechanisms, more precisely, refraction, reflection and diffraction.

The interactions between a light beam and a given particle depend mainly on the ratio between the particle size and the wavelength of the beam of light. In order to precise the domain of each phenomenon, a size parameter α is defined [8]:

$$\alpha = \pi d/\lambda$$

where d is the particle diameter and λ the wavelength.

Figure 4 gives a schematic presentation of the different mechanism involved in light diffusion.

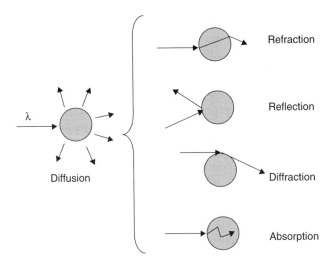

FIGURE 4. *Main mechanisms of light diffusion.*

According to the value of the parameter α, and considering the light sources usually used (UV, visible, near IR), three domains are considered [8]:

- Particles greatly submicronic ($\alpha < 0.3$). For $\alpha < 0.3$, the optical model of the Rayleigh diffusion is used. Particles diffuse light as much forward as backward.
- Particle size superior to several microns ($\alpha > 30$). Laws of optical geometry and diffraction are used. In this case, the light is diffracted, meaning that the diffusion is mainly concentrated in front of the particles.
- Micronic particles ($0.3 < \alpha < 30$). This intermediate domain corresponds to the validity limit of the last models (diffusion and diffraction), because diffusion is also influenced by reflection, refraction and absorption phenomena. These can be taken into account by the complex theory of Lorenz–Mie.

There are, however, some limitations to this presentation [9]. The diffusion is only valid for spherical particles and single diffusion. The shape of the particles (spherical in Mie model), has a strong influence on optical properties [10], as well as the orientation of particles [11]. Thus, other approaches have been proposed for nonspherical particles [12,13]. However, extinction observations are used for estimating the size of scattering particles in granulometric methods. Finally, if the particle size is comparable to the wavelength of light, the extinction will depend on the particle's shape. For aggregate particles with a size comparable to the wavelength, the spectral dependence of extinction efficiency is less steep than that for equivalent spheres, and its maximum is shifted to larger size parameters, i.e. smaller wavelengths [14].

There are very few studies concerning the UV-visible responses of particles. One study [15] on organic pigments in aqueous dispersions has clearly shown that the optical response is the result of absorption and scattering of light by the pigment particles. The intensity of the two phenomena (absorption and scattering) is used to determine the dispersion degree of organic pigments. Another study [16] has proposed a model for the quantitative interpretation of UV-visible spectra of microorganism suspensions (with some specific absorption bands). The model is namely based on light scattering theory and spectral deconvolution techniques. Finally, Thomas *et al.* [17,18] have proposed reference spectra of suspended solids and colloids to model the interferences of UV spectra of wastewater and natural waters.

3.2. Methods for the study of heterogeneous fractions

A variety of techniques are available for the characterisation of particle size distributions but, because of the large size distribution of solids in water and wastewater, no single analytical method can be used. Besides their physico-chemical properties (stability, settleability, etc.), heterogeneous fractions have interesting optical properties. This is the reason why optical methods take an important place among granulometric methods (Table 5).

Microscopic analysis, sieves or membrane techniques are also used for the separation of solids in waters [6]. These optical, physical or electrical techniques are rather easy to use, but need specific and often expensive instruments.

The principles of granulometric methods are more often based on the interaction between spherical particles and light or other physical resistance. This constraint is not a problem for industrial application where the suspension granulometry is generally

TABLE 5. *Main granulometric methods [19]*

Optical methods	Methods based on the interaction between fluid and particles	Electrical method
– Light scattering – Laser diffusion – Light absorption – X-ray absorption Laser diffraction – Light blockage	– Permeametry – Sedimentation in a liquid – Sedimentation in a gas – Centrifugation	– Electrical resistance variation

well controlled. Unfortunately, wastewater suspensions are very variable in nature, size and properties. This is the reason why the use of granulometric methods is very limited.

3.3. UV-visible responses of mineral suspensions

A recent study has shown the relation between UV-visible response and size of some mineral suspensions in water [9]. After filtration of commercial suspensions of talc, kaolin and carbonate (Table 6), the UV-visible spectra of the different granulometric fractions are normalised (see Chapter 2) and compared (Table 7).

- For particles whose diameter is larger than 10 μm, the diffusion domain is the diffraction one. This phenomenon is characterised by UV-visible spectra with absorbance values slightly dependent on wavelength (diffraction and Mie/diffraction domain), which are almost independent from the wavelength. The ratio between absorbance values at 200 and 800 nm is about 2.
- For TSS (>1.2 μm), the diffusion domains are both Mie and diffraction ones. TSS presents the same optical response as particles whose size is greater than 10 μm even if a marked slope is noticed for kaolin. The ratio between absorbance values at 200 and 800 nm is about 2, except for kaolin, because of organic contaminant.
- For colloids, the diffusion domain is the Mie one. UV-visible response depends strongly on wavelength. The ratio between absorbance values at 200 and 800 nm ranges from 10 to 24.

TABLE 6. *Main granulometric characteristics of the studied suspensions [9]*

Suspension	Main modes
Talc	10 μm
Kaolin (slurry)	0.6 and 2 μm
Carbonate (slurry)	0.5 and 2 μm
Kaolin (powder)	5 and 50 μm
Carbonate (powder)	10 μm (broad granulometric spectrum)

TABLE 7. *UV-visible responses (normalised spectra) of particles according to their size [9]*

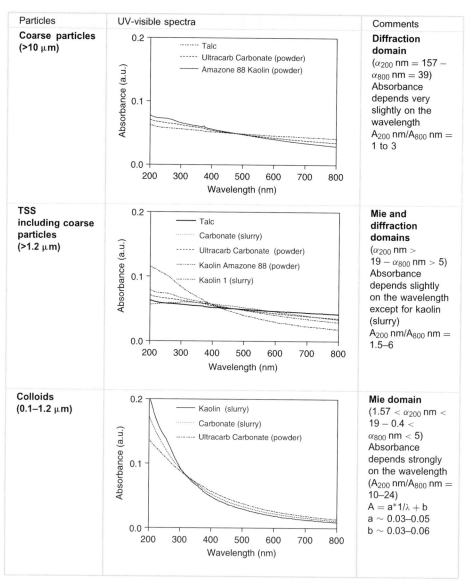

Particles	UV-visible spectra	Comments
Coarse particles (>10 μm)		**Diffraction domain** (α_{200} nm = 157 − α_{800} nm = 39) Absorbance depends very slightly on the wavelength A_{200} nm/A_{800} nm = 1 to 3
TSS including coarse particles (>1.2 μm)		**Mie and diffraction domains** (α_{200} nm > 19 − α_{800} nm > 5) Absorbance depends slightly on the wavelength except for kaolin (slurry) A_{200} nm/A_{800} nm = 1.5–6
Colloids (0.1–1.2 μm)		**Mie domain** (1.57 < α_{200} nm < 19 − 0.4 < α_{800} nm < 5) Absorbance depends strongly on the wavelength (A_{200} nm/A_{800} nm = 10–24) $A = a*1/\lambda + b$ a ~ 0.03–0.05 b ~ 0.03–0.06

Moreover, a linear relation between normalised absorbance values and the reciprocal wavelength (Fig. 5) can be proposed to model the UV response of colloids

- $A = 35.3* 1/\lambda − 0.032$ $R^2 = 0.9976$ (carbonate Ultracarb, powder)
- $A = 42.6* 1/\lambda − 0.048$ $R^2 = 0.9916$ (carbonate, slurry)
- $A = 49.4* 1/\lambda − 0.064$ $R^2 = 0.973$ (kaolin, slurry)

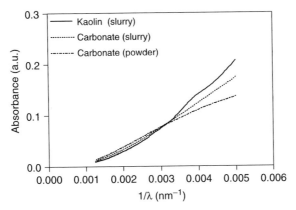

FIGURE 5. *Relation between absorbance values and 1/wavelength, for commercial suspensions [9].*

The difference observed between UV spectra of particles in slurry form and in powder form can mainly be explained by the presence of organic compound(s) used in the formulation that may absorb in the UV region (particularly for kaolin). For the range of 200–300 nm ($1/\lambda$ from 0.003 to 0.005), carbonate powder spectrum deviates negatively and kaolin deviates positively from linearity. Further validation should be made to check the observed linearity.

3.4. UV responses of wastewater

UV absorption spectra of wastewater are not easy to understand in view of their featureless shapes, partly due to the effect of suspended matters. The latter are very heterogeneous and responsible for diffuse absorbance more intense in the UV region than in visible. Actually, UV responses of particulate and colloidal matter are often the result of both chemical and physical responses related to their nature (organic for a great part and able to adsorb soluble compounds such as surfactants). For example, in case of slate particles (mineral) of a few micrometers, the spectrum is flat. The absorbance is uniform and higher for smaller particles. The chemical absorbance related to the presence of suspended solids of organic nature seems to emphasise spectrum slope and to create shoulders (Fig. 6).

The role of settleable matter and supracolloids in the absorbance of suspended solids of wastewater has been studied more precisely. A simple experience has been carried out. Raw wastewater has been introduced into an Imhoff cone over 1 h. Figure 7 presents the settling device used for the separation of four granulometric fractions. These fractions have been analyzed both with laser granulometry (laser diffraction) and with UV spectrophotometry (deconvolution method, see Chapter 2).

With laser diffraction technology, the particle size distribution is determined on the basis of the scattering monochromatic light (at 750 nm). The measurement of particle diffraction pattern is characteristic of particle size in the range of 0.4 to 2000 μm of particle diameters [20].

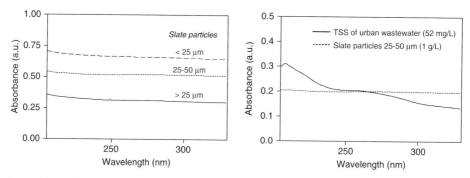

FIGURE 6. *Diffusion of UV light (left: slate particles of different size distributions, right: normalised spectra of slate particles and urban TSS).*

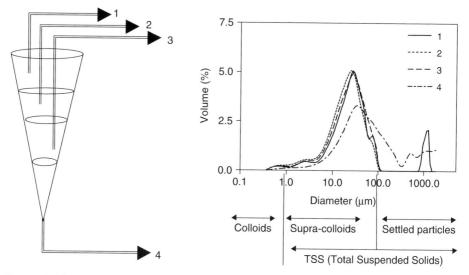

FIGURE 7. *Granulometric size distribution study (left: separation device, right: laser granulometry results) [20].*

Granulometric analysis of fractions 1, 2, 3 (Fig. 6) clearly shows that particles are mainly supracolloids, between 1 to 100 μm, with a mode value close to 30 μm. The fraction 4 (at the bottom of the Imhoff cone) presents the widest range size with particle larger than 100 μm, and a multi-modal distribution. After 1 h of settling, the separation of settleable matter is achieved, as fractions 1 to 3 do not contain particles of size above 100 μm, even if, in the first fraction, some millimetric floating particles are present. One can note that no colloidal population is detected by laser granulometry.

Figure 8 shows that UV spectra of the fractions 1, 2 and 3 are superposed, confirming the granulometric results. Fraction 4 presents a higher absorbance due to the presence of suspended solids. Indeed, the quantity of colloids (quantified by coefficient contribution calculation) appears to be the same for the four fractions.

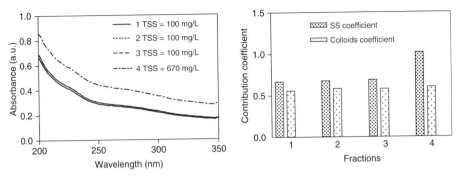

F<small>IGURE</small> 8. *UV spectra (right) and repartition of SS and colloids contribution (left) of four fractions (UV estimated coefficients by deconvolution method).*

The absorbance is not proportional to the concentration, and can be explained by the fact that settleable particles, the sizes of which are above 100 μm, diffuse the light slightly by diffraction. Thus, light diffusion by TSS is mainly due to the presence of supracolloids.

The normalisation of the set of spectra (Fig. 9) allows comparing the quality of each fraction. The slope break of the UV spectrum of fraction 4, around 240 nm, is softened, confirming the presence of large particles [21]. If the discrimination between supracolloids and settleable matter is not possible by UV spectrophotometry, the laser granulometry results confirm the major role of supracolloids in diffuse absorbance.

Thus, it appears clear that the advantage of using of UV spectrophotometry for the study of suspended solids is that the UV response integrates the effect of the different solid classes and gives an average signal of the solid mixture.

Before considering the quantification of total suspended solids of water and wastewater, the relevance of UV spectrophotometry should be investigated for the study of different types of wastewater. The comparison of the behaviour towards UV light of various size ranges studied for different water and wastewater types is reported in Table 8.

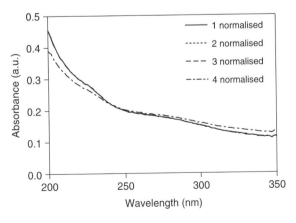

F<small>IGURE</small> 9. *Normalised UV spectra of Fig. 8.*

TABLE 8. *Spectra of size range fractions according to the water type (normalised spectra)* [22]

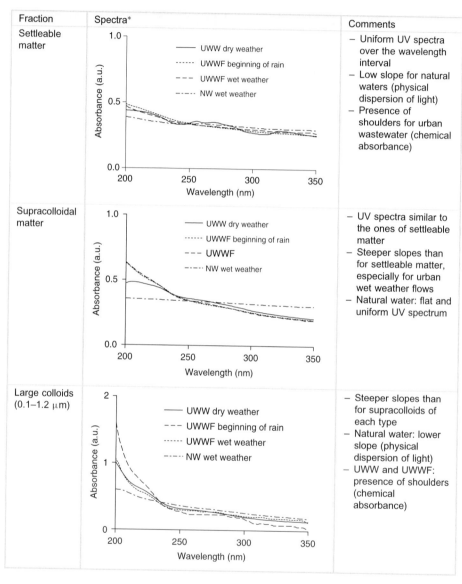

Fraction	Spectra*	Comments
Settleable matter		– Uniform UV spectra over the wavelength interval – Low slope for natural waters (physical dispersion of light) – Presence of shoulders for urban wastewater (chemical absorbance)
Supracolloidal matter		– UV spectra similar to the ones of settleable matter – Steeper slopes than for settleable matter, especially for urban wet weather flows – Natural water: flat and uniform UV spectrum
Large colloids (0.1–1.2 μm)		– Steeper slopes than for supracolloids of each type – Natural water: lower slope (physical dispersion of light) – UWW and UWWF: presence of shoulders (chemical absorbance)

TABLE 8. *cont'd*

Fraction	Spectra*	Comments
Fine colloids (including some macro-molecules), molecular weight > 10 kD and size < 0.1 μm	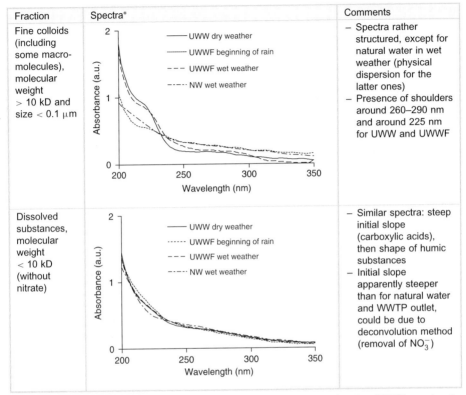	– Spectra rather structured, except for natural water in wet weather (physical dispersion for the latter ones) – Presence of shoulders around 260–290 nm and around 225 nm for UWW and UWWF
Dissolved substances, molecular weight < 10 kD (without nitrate)		– Similar spectra: steep initial slope (carboxylic acids), then shape of humic substances – Initial slope apparently steeper than for natural water and WWTP outlet, could be due to deconvolution method (removal of NO_3^-)

*UWW: urban wastewater; UWWF: urban wet weather flow; NW: natural water; WWTP: wastewater treatment plant.

Some observations related to UV spectra shape corresponding to different types of water and wastewater can be expressed [22]:

- The smaller the particles are, the higher is the absorbance value.
- The spectrum of settleable matter is generally flat. The UV spectrum of suspended solids of natural water, mainly of mineral nature, is essentially related to the diffusion of light.
- For smaller particles, the spectrum is close to the one of the reference spectrum of dissolved substances (see Chapter 4). Chemical absorbance induces a shoulder around 225 nm, often associated with the presence of surfactants in wastewater [23]. This shoulder is especially marked on UV spectra of fine colloids in raw wastewater. This tends to show the general affinity of soluble compounds for fine particles and confirms that surface phenomena are more important for finer particles than for the larger ones. This behaviour can also be explained by the presence of macromolecules such as humic-like substances.

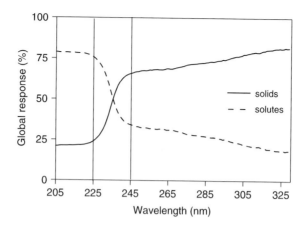

F<small>IGURE</small> 10. *Theoretical contribution of solids and solutes in UV response of water.*

A comparison can be made between Table 7 and Table 6 (mineral suspensions). In both cases, the presence of coarse particles is characterised by a diffusion spectrum with a relatively flat shape due to the very light dependence of absorbance versus wavelength (diffraction). On the contrary, the presence of colloids is responsible for a high variation of absorbance with wavelength (Mie diffusion). The main difference between the two sets of spectra is the presence of structured elements (i.e. shoulders) on the spectra of wastewater, related to the chemical absorption of organic compounds bound to the particles (i.e. surfactants).

It is also interesting to study the global response of the main phases of wastewater. As the semi-deterministic method of spectra exploitation uses reference spectra (see Chapter 2), the contribution of main phases (dissolved and solids) can be calculated at each wavelength from the corresponding absorbance values. Assuming that a theoretical spectrum is the sum of the reference spectra (each coefficient contribution being equal to 1), the contribution of the dissolved phase can thus be calculated at each wavelength from the sum of organic matrix and surfactants and nitrate divided by the sum of absorbance of reference spectra. The contribution of the solid phase is calculated from the sum of suspended solids and colloids spectra.

Figure 10 displays the evolution of this global response and shows that the dissolved fraction is predominant with regard to the UV absorbance below 225 nm, while the solid one has a more important contribution above 245 nm. Between the two phases, the same theoretical influence on UV responses can be noted. This observation could explain that, in numerous cases, UV spectra present a break in their shape at around 235–240 nm.

4. TSS ESTIMATION

TSS measurement can be carried out by gravimetry after separation by filtration or centrifugation, according to standard methods [5,24]. For online measurement, these techniques cannot be applied without expensive devices. This is the reason why optical methods are often used, especially in industry. Among the different systems, turbidimetry, which is widely used for water quality control as an alternative method for suspended solids, is probably the simplest way to get a coarse estimation of both suspended solids and a part of the colloidal fraction.

4.1. Turbidimetry

Among the various available techniques for turbidity measurement, the main one called nephelometry is based on the reflection of light induced by particles in water. The angle between the source (a red LED, for example) and the detector is 90°. Another principle used for turbidimetry is back scattering, sometimes used for higher concentrations of solids. The intensity of the detected light is a function of the number of particles and their apparent size. Turbidity measurement depends on concentration, size and surface properties of particles [25,26]. It mainly considers the colloidal fraction [27], which limits its use to wastewater of constant quality. Turbidity is thus affected by both suspended solids and colloidal matter of a settled sample, and by the distribution between colloids and supracolloids. The latter point implies that the estimation of suspended solid concentration is only possible if the solids to be characterised have a very simple granular spectrum with a single class of particles. In this case, a calibration between turbidity and suspended solid concentration will roughly lead to the estimation of TSS. Nevertheless, a broad granular spectrum can sometimes have a rather constant distribution, and a relation can thus be established between turbidity and suspended solids also. Such a relation needs to be checked regularly. This can be the case for raw wastewater coming from a separate sewer system at the inlet of a treatment plant.

On the contrary, it is very difficult to find a useful relation between these two parameters for treated wastewater since, on one hand, the calibration range is weak, and on the other hand, the granular spectrum is related to one functioning point of the treatment plant. This consideration excludes the use of turbidity for the control of chemically treated wastewater. However, a method to extract particle concentration and characteristic particle size from turbidimetre readings has been developed [28]. The method requires a turbidimeter capable of measuring the forward (12°) and sideways (90°) scattered light simultaneously. Tested on the calibration of filter aids, an industrial pigment and yeast, the method is limited to the range of particle concentration (1–200 ppm) and characteristic particle size (1–100 μm) used in the calibration. A second limitation arises from the fact that the other parameters influencing scattering (optical properties, particle shape and porosity, width of the size distribution) are not considered. In conclusion, the estimation of suspended solids from turbidity measurement is hazardous and has to be considered only for water and wastewater of constant quality with regard to the particulate fraction.

Table 9 presents the potential application of turbidimetry and other optical techniques such as absorptiometry for high concentration. This type of measurement can be proposed for the estimation of TSS, according to the nature of the sample. In any case, the calibration step must be carefully established and regularly checked.

Before considering the use of UV spectrophotometry, the use of simple absorptiometry (generally around 700–800 nm) for suspended solid measurement should be mentioned. Even if this technique presents the same drawbacks as nephelometry, it can be used when the colloidal fraction is negligible, e.g. for the estimation of sludge concentration. A better result is obtained if the sample is first grounded by ultrasonic or mechanical means [29].

TABLE 9. *Application of optical measurement for the estimation of suspended solids*

Sample	Concentration range	Size characteristics	Usual method	Estimation quality
Raw sewage	100–600 mg/L	Very wide size range	Turbidimetry (back scattering)	+/−
Treated water (biological treatment)	5–100 mg/L	Homogeneous solids (bacterial flocs)	Turbidimetry (nephelometry)	+/−
Treated water (chemical treatment)	5–100 mg/L	Wide size range (colloids and flocs)	Turbidimetry (nephelometry)	−
Activated sludge	1–10 g/L	Homogeneous solids (bacterial flocs)	Absorptiometry	+

4.2. UV estimation of TSS

UV estimation has been carried out using the semi-deterministic method of UV spectra exploitation (see Chapter 2). Because of the complexity of the light diffusion phenomena according to the type of water or wastewater, TSS estimation depends on either SS coefficient alone or both colloids and SS coefficients.

The following relation is used for TSS estimation after UV spectrum deconvolution:

$$[TSS] = \alpha_{SS} \cdot C_{SS} + \alpha_{Coll} \cdot C_{Coll} + r$$

where α_{SS} and α_{Coll} are, respectively, the values of the contribution coefficients of the reference spectra corresponding to suspended solids and colloids, C_{SS} and C_{Coll}, the equivalent concentrations of TSS (statistically determined) associated with these reference spectra, and r is the error on the computation of the parameter value (see Chapter 2).

Figure 11 presents the results of a comparison between the UV method and a standard one (EN 872) obtained from urban treated wastewater (physico-chemical treatment).

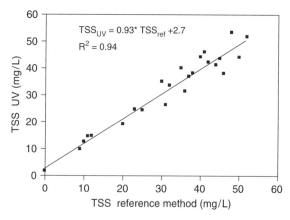

FIGURE 11. *Comparison between reference and alternative method (urban wastewater treated by physico-chemical process).*

TABLE 10. *UV estimation of TSS*

Origin	Concentration range mg/L	Determination coefficient (R^2)
Raw urban wastewater	5–400	0.89
Treated urban wastewater (biological)	5–100	0.91
Treated urban wastewater (physico-chemical)	5–50	0.94
Industrial raw effluent	5–100	0.77
Industrial treated effluent (biological)	5–50	0.83

This method has been applied for several wastewater types and gives relatively good results (Table 10) except for raw industrial wastewater where the quality is highly variable [17,18]. The application of the UV method on natural water is more difficult because of the solids' nature and of the concentration range (from a few mg/L to several g/L in case of floods).

References

1. C.R. O'Melia, M.W. Hahn, C.T. Chen, *Wat. Sci. Technol.*, 36 (1997) 119.
2. A.D. Levine, G. Tchobanoglous, T. Asano, *J. Water Pollution Control Fed.*, 57 (1985) 805.
3. J.L. Balmat, *Sewage Ind. Waste*, 29 (1957) 757.
4. H. Heukelekian, J.L. Balmat, *Sewage Ind. Wastes*, 31 (1959) 413.
5. *Standard Methods for the Examination of Water and Wastewater*, 2450 F, Settleable Solids, 20th Edition, American Public Health Association, Washington, USA (1998).
6. A.D. Levine, G. Tchobanoglous, T. Asano, *Wat. Res.*, 25 (1991) 911.
7. S. Görög, *Ultraviolet-Visible Spectrophotometry in Pharmaceutical Analysis.* CRC Press, Boca Raton, New York-London-Tokyo, (1995).
8. T. Allen, *Particle Size Measurement*, 4th Edition, Chapman and Hall, (1990).
9. C. Berho, M.-F. Pouet, S. Bayle, N. Azema, O. Thomas, *Coll. Surf. A*, 248 (2004) 9.
10. A.R. Jones, *Prog. Energy Combust. Sci.*, 25 (1999) 1.
11. M. Naito, O. Hayakawa, K. Nakahira, H. Mori, J. Tsubaki, *Powder Technol.*, 100 (1998) 52.
12. K. Chamaillard, S.G. Jennings, C. Kleefeld, D. Ceburnis, Y.J. Yoon, *J. Quant. Spectrosc. Radiat. Transf.*, 79–80 (2003) 577.
13. V.M. Rysakov, *J. Quant. Spectrosc. Radiat. Transf.*, 87 (2004) 261.
14. E.V. Petrova, K. Jockers, N.N. Kiselev, *Icarus*, 148 (2000) 526.
15. J.M. Fu, Y. Li, J.L Guo, *J. Coll. Interface sci.*, 202 (1998) 445.
16. C.E. Alupoaei, J.A. Olivares, L.H. Garcia-Rubio, *Biosens. Bioelectron.*, 19 (2004) 893.
17. O. Thomas, F. Théraulaz, M. Domeizel, C. Massiani, *Environ. Technol.*, 14 (1993) 1187.
18. O. Thomas, F. Théraulaz, C. Agnel, S. Suryani, *Environ. Technol.*, 17 (1996) 251.
19. IFTS, Characterization of solids in wastewater networks, *Report*, Institut de la Filtration et des Techniques Séparatives, Agen, France (1989).
20. N. Azema, M-F. Pouet, C. Berho, O. Thomas, *Coll. Surf. A*, 204 (2002) 131.
21. S. Vaillant S., M-F Pouet, O. Thomas, *Urban Water*, 4 (2002) 273.
22. S. Vaillant, *Organic Matter of Urban Wastewater: Characterisation and Evolution.* PhD thesis, University of Pau and Pays de l'Adour, France (2000).
23. F. Théraulaz, L. Djellal, O. Thomas, *Tenside Surf. Det.*, 33 (1996) 447.
24. AFNOR EN 872, *Suspended Solids Measurement-Method by Fiberglass Filtration*, Association francaise de normalisation, France, 1996.
25. D.S. Bhargava, K. Rajagopal, D.W. Mariam, *Indian J. Engin. Mat. Sci.*, 2 (1995) 217.

26. J-L. Bertrand-Krajewski, D. Laplace, C. Joannis, G. Chebbo, *Pollutant Measurement by Optical Methods, In: Measurement in Urban Hydrology and Sanitation*, B. Chocat (ed.), Lavoisier Tec & Doc, Paris, 1997.
27. O. Thomas, *Pollution by Particles, In: Urban Wastewater Metrology*, O. Thomas (ed.), Lavoisier Tec & Doc, Paris, 1995.
28. H.H. Kleizen, A.B. de Putter, M. van der Beek, S.J. Huynink, *Filtrat. Sep.*, 32 (1995) 897.
29. J.P. Denat, O. Thomas, M. Roulier, M. Martin-Bouyer, *Trib. Cebedeau*, 33 (1980) 73.

UV-Visible Spectrophotometry of Water and Wastewater
O. Thomas and C. Burgess (Eds.)

163

CHAPTER 7

Natural Water

M.-F. Pouet[a], F. Theraulaz[b], V. Mesnage[c], O. Thomas[a]

[a]Observatoire de l'Environnement et du Développement Durable, Université de Sherbrooke, Sherbrooke, Québec, J1K 2R1, Canada; [b]Laboratoire Chimie et Environnement, Université de Provence, 3 place V. Hugo, 13331 Marseille Cedex, France; [c]Département de Géologie, UMR CNRS 6143, Université de Rouen, 76821 Mont-Saint Aignan, France

1. INTRODUCTION

The composition of water and wastewater depends, on one hand, on the hydrogeochemical context responsible for the mineral matrix, and on the other hand, on the different inputs such as storm runoff or discharge of treated wastewater that carryies pollution loads (Fig. 1). From a physical point of view, natural water is characterised by the presence of dissolved constituents (of mineral or organic forms), including gas (oxygen, carbon dioxide, etc.) and heterogeneous fractions (suspended solids with particles greater than 1 μm, or colloids, between 1 and 0.1 μm). This composition was discussed in the previous chapter.

The expression "natural water" refers to water from a large array of sources. From drinking water, including tap and mineral water, to seawater, a large scale of mineralisation is covered, from few milligrams per litre to several tenths of grams of dissolved solids per litre. In between, groundwater and surface water, including rivers and lakes, ponds and wetland, can be found. Except for these last media, either from natural (ponds and wetlands) or anthropogenic (polluted water) origin, the concentration of organic compounds is generally much lower than for inorganic constituents, with only few milligrams per litre or less. Natural organic matter (NOM) is composed of all organic compounds mainly issued from the degradation of vegetal or animal biomass. It includes a lot of degradation by-products, from leaves and wood decomposition, but also humic-like substances. Anthropogenic organic matter (AOM) occurs when wastewater, generally treated, is discharged in water. It is composed of biodegradable organic compounds from domestic activities (proteins, carbohydrates, fats, surfactants, etc.) but can be completed by biorefractory organic compounds. As a result of biodegradation of AOM in water, under microorganisms and oxygen, organic and mineral compounds, such as carboxylic acids, phosphate and nitrate, are produced. These last are considered as nutrients or as inorganic mineral matter (IMM). Figure 2 presents the evolution of organic matter in water.

Before considering the use of UV spectrophotometry for natural water quality control, it is important to underline that the visual observation of the aqueous media (river, lake, wetland, etc.) can be of great interest. Actually, several indicators can reveal the level of pollution (present or past) of water.

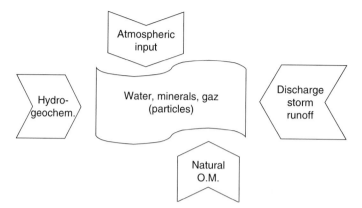

FIGURE 1. *Factors influencing the composition of natural water.*

River water may have deposed some solid wastes on the banks (plastics, glass, metallic wastes, etc.). The aspect and colour of interface with water can also be significant: brown mud or bacteria aggregates related to an existing pollution, green to a past one (presence of nutrients), orange to the presence of iron in petlands, etc. Moreover, the absence of plants on the banks can also be a sign of pollution (except for rocky site).

Water bodies can be turbid, slightly brown or green, with aquatic plants such as algae or macrophytes (duckweed, reeds, etc.) in case of eutrophication. Living species such as fishes or small animals (proto and metazoaires) can be associated with a specific pollution level. Further biological consideration can be found in literature.

2. SIGNIFICANCE OF UV SPECTRA OF NATURAL WATER

The use of UV spectrophotometry for the characterisation of natural water quality has led to several quantitative procedures, and the main ones are presented in Table 1. The first UV measurement for an environmental parameter (nitrate ion) was proposed 50 years ago [10]. Based on the simple exploitation of UV spectrum, the measurement of absorbance at one (or two) wavelength has also been applied for other parameters such as COD and followed by multiwavelength procedures (see Chapter 2). This quick literature synthesis shows that only few parameters or compounds can be studied from

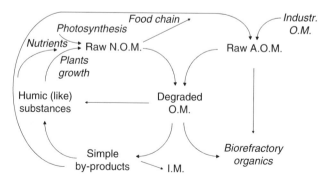

FIGURE 2. *Evolution of organic matter in water.*

TABLE 1. *Use of UV spectrophotometry for natural water quality study*

Parameter	Wavelength (nm)	Procedure	Application	Authors
Natural organic matter	220	Absorptiometry Absorbance ratio	Sea Polluted rivers	Ogura et al. [1] Dobbs et al. [2]
	225/250/275/350		Sea and rivers	Foster [3]
COD	254	Absorptiometry	Surface water, Effluents	Briggs et al. [4] Mrkva [5,6]
COD, DOC	205–330	Deconvolution	Surface water, effluents	Thomas et al. [7–9]
Nitrate	210/275	Absorptiometry	Surface water	Hoather [10]
	205–250	Polynomial mod.	Surface water, effluents	Thomas et al. [11]
	205–330	Deconvolution	Surface water, effluents	Thomas et al. [7–9]
Humic substances	250 270	Absorptiometry	Lakes	DeHaan [12] Buffle[13]
Chromium(VI)	250–450	Polynomial mod. Deconvolution	Surface water, effluents	Thomas et al. [14] El Khorassani [15]
Trace organics	250–350	Multicomponents	Surface water	Maier [16]

UV spectra of water. Thus, it is possible to classify UV spectra of natural water into four groups, including suspended solids and chloride response (Fig. 3).

The first type of spectrum is characterised by the presence of nitrate. As nitrate is the most stable form of nitrogen in water, resulting from the oxidation of all other dissolved N compounds (nitrite, ammonia or organic nitrogen), the majority of natural water contains nitrate. Taking into account the UV spectrum shape of nitrate (see also Chapter 5), the presence of nitrate is easy to recognise if the concentration is greater than 5 mg/L (of $N–NO_3^-$). Even in presence of other compounds constituting the organic matter matrix, a convex form in the 205–220 nm region is, most of the time, related to the presence of nitrate.

The second type of UV spectrum often encountered with natural water is characterised by a general decreasing concave shape on a wide range of wavelengths, with a slight shoulder at around 260–270 nm. The absorbance values are rarely close to zero, even in the higher wavelengths. This is due to the presence of NOM and, more precisely, humic substances (HS) or humic-like substances (HLS) composed of humic acids, fulvic acids and related substances. Humic acids have the property of precipitating in acid solution (pH < 2), in contrast to fulvic acids, representing the major part of HS (between 80 and 85% in weight), which are more soluble and extractable in basic condition (pH > 10). These compounds correspond to the major part of NOM, and of DOC (between 50 and 70%, on average). Even if some recent study advances a less complex composition [17], HS or HLS can be considered as macromolecules of high molecular weight, including several chromophore groups due to unsaturated binding sites susceptible to absorb significantly in the UV, as phenolic or carboxylic functions, for example. This is the reason why no specific shape of UV spectrum, such as peaks or marked shoulders, is associated with this organic matrix. The resulting shape is rather constant and seems to result from the overlapping of a lot of spectra of more simple organic compounds. The slight shoulder around 260–270 nm, which can be accompanied by a 220–230 nm one, is due to a higher percentage of phenolic-type chromophores associated with some HLS of ponds or wetlands. In some cases, it is possible to associate a particular concave absorbance evolution between 200 and 220 nm due to the presence of carboxylic compounds.

M.-F. Pouet et al.

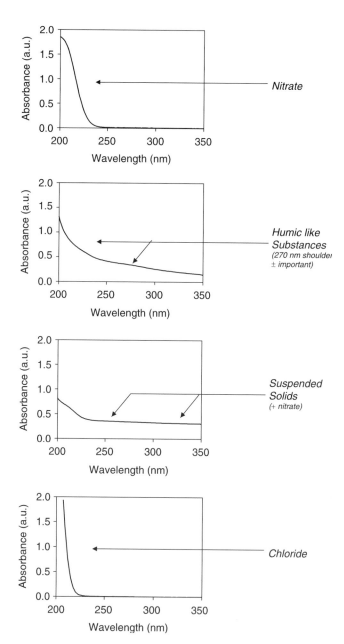

FIGURE 3. Main types of UV spectra of natural water.

The third type of UV spectrum of natural water is rather simple to recognise as the major part of the spectrum is more or less linear and horizontal, with absorbance values rather high. This is due to the physical diffuse absorption response of suspended solids, as explained in Chapter 6. Depending on the nature of solids and on the presence of high colloid concentration, the absorption level and the general mean slope of the spectrum in the 250–350 nm region can vary. Samples must often be diluted in this case, in order to prevent the saturation of absorbance values. Considering that natural water often contains nitrate, the resulting spectrum obviously shows the convex characteristic signal.

The fourth main type of UV spectrum of natural water is the one in seawater with a high content of chloride. As shown in Chapter 5, chloride is responsible for an absorption wall below 220 nm, for high concentration. As no other constituent is present in high concentrations in seawater, the spectrum is generally close to zero for wavelengths greater than 225 nm.

In addition to these four main types of UV spectra of natural water, many more can be encountered. For memory, when the flow measurement of river, using dichromate, was formerly authorised, typical spectra of hexavalent chromium were obtained (see Chapter 9). In case of pollution, the UV spectrum shape is obviously dependent on the pollutant nature and concentration. In all cases, if the UV spectrum of natural water is flat and close to zero, the pollution probability is very low. On the contrary, a more or less important UV spectrum is always related to the presence of dissolved compounds or suspended solids.

Several real case studies are presented in the next sections, on different water bodies (river, lake, groundwater, etc.), showing the interest of UV spectrophotometry for qualitative (quality characterisation and evolution) and quantitative (concentration and parameters estimation) applications.

3. QUALITY OF NATURAL WATER

3.1. Study of water quality variation along a river

The first example is the study of the water quality of the Gardon de Saint-Jean River located in the South of France in the Cevennes region. This 50-km-long river is of the Mediterranean type, with a very low flow rate in summer and frequent flash floods the rest of the time. From the pollution point of view, this river is characterised by the existence of two main villages (of about 2000–5000 inhabitants), with two wastewater treatment plants, but it is also the heart of a tourist area with an estimated overpopulation of 50,000 people in summertime.

A sampling campaign has been carried out in summer 2000, with nine points located along the river (Fig. 4). For each station, the UV spectrum and the measurement of dissolved oxygen, conductivity and pH have been realised on fresh samples. A field-portable UV spectrophotometer (Pastel UV from Secomam) and handheld instruments for the other parameters have been used.

The evolution of the physico-chemical parameters is presented in Fig. 5. Close to the source, the dilution of the river can be noted with water of low mineralisation. An increase of dissolved oxygen expressed as saturation percentage is observed, as well as an increase in the pH values downstream.

Figure 6 displays the evolution of UV spectra along the river. As usual, nitrate ion is present in most of the samples, and the organic matter content is rather low. The study of normalised spectra (see Chapter 2) shows the existence of two isosbestic points related

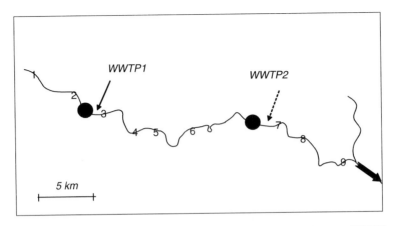

FIGURE 4. *Sampling location on the Gardon de Saint-Jean River, France (WWTP: wastewater treatment plant).*

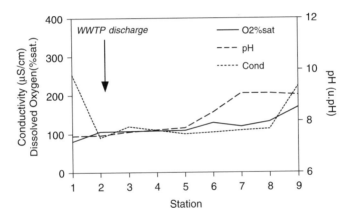

FIGURE 5. *Evolution of physico-chemical parameters (average water temperature = 25°C).*

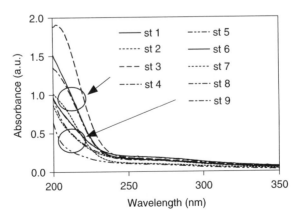

FIGURE 6. *Evolution of UV spectra.*

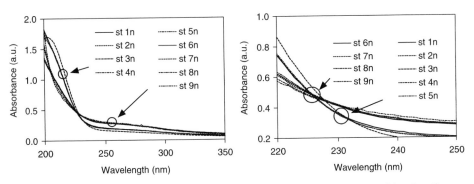

FIGURE 7. *Isosbestic point revelation after normalisation (at right, zoom of isosbestic point).*

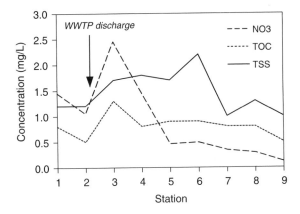

FIGURE 8. *Evolution of UV-estimated pollution parameters.*

to two groups of spectra (Fig. 7). The first group concerns the upstream points under the influence of the discharge of the main wastewater treatment plant (WWTP1). The second group of spectra is related to the consumption of organic matter and nitrate in the second part of the river. This phenomenon is confirmed in Fig. 8, by the evolution of pollution parameters, TOC, TSS and NO_3^-, estimated by UV (see Chapters 4 and 5).

The conclusion of this experiment is that, even if the pollution level of the river is low, due to high performances of the wastewater treatment plants, the use of UV spectrophotometry simplifies the field (and laboratory) work by bringing complementary information on the evolution of water quality.

3.2. Rain influence on river water quality

Another example is related to the study of the influence of rain on river water quality. As shown in Fig. 3, the presence of suspended solids leads to a typical shape of spectra. Figure 9 shows spectra of different rivers of the South of France during floods,

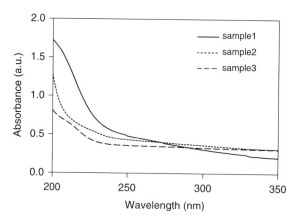

FIGURE 9. *Spectra of rivers water during floods.*

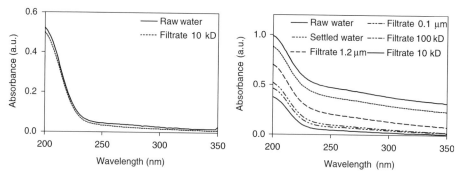

FIGURE 10. *Fractionation of natural water sampled in dry weather conditions (left) and under rain conditions (right) [18].*

in summer 2001. Spectra of samples 1 and 3 are different for their nitrate concentration and solids nature and origin, while sample 2 probably contains more NOM than the two others. In order to study the granulometric characteristics of solids, a fractionation experiment is carried out on sample 1.

Figure 10 presents the fractionation of natural water sampled during dry weather and wet weather [18].

The river water during dry weather is characterised by the presence of nitrate (around 3 mg/L) and a low concentration of organic matter (around 1.8 mg/L of TOC). The filtration of the water is very easy (the TSS concentration being close to zero), and the spectrum of the filtrate (after filtration at 10 kD, i.e. about 0.001 μm) is very similar to that of raw water, indicating a very low quantity of colloids.

During wet weather, the river water sample contains a TSS concentration of 276 mg/L (measured by a reference method), principally of mineral composition as organic fraction represents only 18%. The size distribution of solids, put in suspension in water by rain flow, is very wide. Settleable, supracolloidal and colloidal solids are present in water. Several filtration steps have been carried out, and the corresponding UV spectra of the

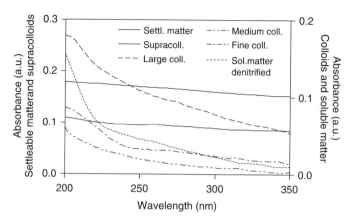

FIGURE 11. *Differential spectra resulting in a wet weather sample fractionation [18].*

filtrates acquired. Figure 11 shows the spectra corresponding to the compounds retained on the filtration membranes, calculated from the difference of the corresponding spectra (for example, settling matter spectrum = raw water spectrum − settled water spectrum). The differences between spectra are important, showing a large distribution of suspended solids and colloids.

The results show that, depending on fractions, several observations can be drawn. The different responses are related to the diffuse absorption response of solids and colloids. The general tendency is that the slope of the spectrum generally decreases as the particle size increases. Spectra corresponding to suspended solids are flat, confirming a mineral nature. Adsorption phenomenon can be seen mainly on fine colloids, and the shape of soluble fractions (artificially denitrified with the deconvolution method; see Chapter 2) is related to the probable presence of humic-like substances (see Chapter 6).

3.3. Study of wetland water quality

Waters from wetlands and lakes may have relatively high concentrations of NOM. A study of surface water quality of wetland has been carried out on one of the Seine estuary, the Hode Marsh. This littoral wetland is located on the northern part of the Seine floodplain in France.

The Hode Marsh is delimited by two different bridges: Tancarville Bridge to the East and the Normandy Bridge to the West. For over a century, the Seine estuary has been highly impacted by human activities: the development of seaport installations and industries, polders for agriculture, the construction of the fluvial navigational canal, etc. All these human actions have severely altered the physical landscape, resulting in the reduction of wetland habitat. Figure 12 presents the location of sampling stations. Sampling stations 1, 2 and 3 are located on mudflats and reed beds. Sample 4 represents the water quality of the Tancarville canal, while sampling stations 5 and 6 are located in wet meadows.

UV spectra of the different samples, taken in summer 2000, are shown on Fig. 13, and the results of physico-chemical parameters are given in Table 2. The results obviously confirm the great difference of water quality between seawater and freshwater.

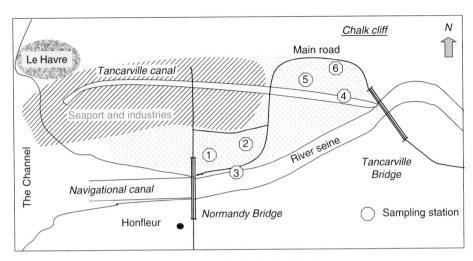

FIGURE 12. *Location of sampling stations in the Hode Marsh.*

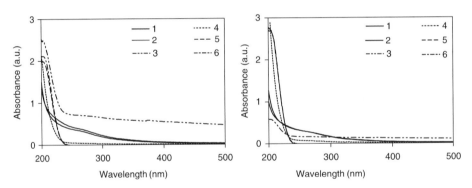

FIGURE 13. *UV spectra of the Hode Marsh samples (Fig. 12). Raw spectra (left) and normalised (right).*

According to the shape of UV spectra and physico-chemical results, different types of waters can be identified. The northern part of the marsh (wet meadows, stations 5 and 6) is only influenced by freshwaters. UV spectra are typical of natural water without organic matter (the absorbance above 240 nm is close to zero). The Gaussian shape around 210 nm indicates the presence of nitrates. UV spectra are superposed, meaning that water quality is the same for these two stations. The conductivity value of 420 μS/cm shows the direct influence of the chalk aquifer.

Sample 4 presents a low concentration of organic matter, giving a residual absorbance on the entire spectrum, except for the shortest wavelengths related to the presence of chloride, the mouth of the canal directly opening into the Channel. This direct influence of salt water explains the high value of conductivity.

UV spectra of samples 1 and 2, similar and superposed, are characteristic of water containing natural organic matter. The diffuse absorbance on the entire spectral domain

TABLE 2. *Physico-chemical characteristics of samples (Fig. 12)*

St.	pH (20°C)	Cond (mS/cm)	Pt (mg.L^{-1})	PO$_4^{3-}$ (mg.L^{-1})	NH$_4^+$ (mg.L^{-1})	COT* (mg.L^{-1})	NO$_3^-$* (mg.L^{-1})
1	8.6	3.02	0.4	0.2	<0.1	9	<0.5
2	7.9	2.11	0.4	0.4	0.7	10	<0.5
3	7.8	1.08	0.5	0.47	0.68	9	<0.5
4	7.9	18.7	—	<0.1	<0.1	<1	<0.5
5	8.9	0.43	<0.1	<0.1	<0.1	<1	33
6	7.5	0.42	0.4	<0.1	<0.1	<1	36

*Estimated by UV.

can be explained by compounds close to humic-like substances. UV spectrum of sample 3 presents an important residual absorbance, showing that this sample is characterised by the presence of suspended solids and organic matter. The presence of nitrate can also be suspected. These results, in addition to chemical characteristics [19], confirm the influence of estuarine water in the south of the marsh. Surface waters are brackish to salt-enriched, with organic matter and nutrients (Table 2).

3.4. Study of lakes water quality

Lakes and ponds may have very variable water quality. Figure 14 shows some examples of UV spectra from samples taken in summer 1998 in French lakes and ponds, mainly located in the Alps. The concentration of organic matter is variable from one water body to another, between 0.8 mg.L^{-1} for Realtor, to 3.4 mg.L^{-1} for Helene and even 5.4 mg.L^{-1} for the studied pond. Nitrate content is very low, with 2.4 mg.L^{-1} as a maximum, for Realtor. This last is characterised by a TSS concentration of 7.4 mg.L^{-1}. After nitrate removal (on UV spectrum; see Chapter 2) and normalisation, two types of water composition appears clearly, as the three corrected spectra seems to be similar. On one hand is mesotrophic or even eutrophic water characterised by the presence of NOM with a majority of HLS, and on the other hand a water of good quality (oligotrophic

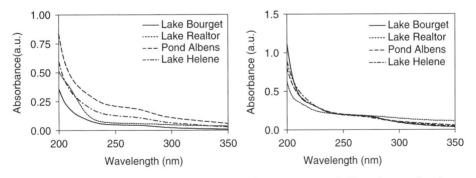

FIGURE 14. *Examples of French lakes and ponds: raw spectra (left) and normalised spectra (right) after nitrate removal.*

medium) with very few organic matter (but a small amount of suspended solids. The considered lake is actually used as a reservoir for a water treatment plant.

Another example is presented in Fig. 15. A total of 21 lakes from Southern Québec, Canada, were sampled in summer 2004, and the corresponding UV spectra are very

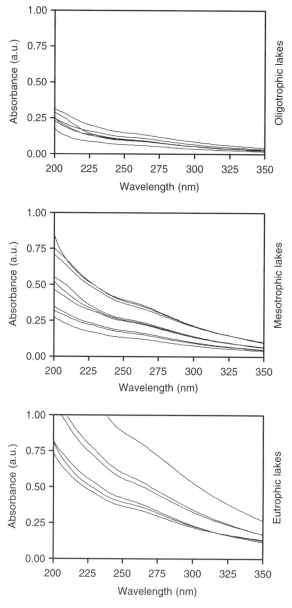

FIGURE 15. *UV spectra of 21 lakes of Southern Québec grouped in function of their trophic state.*

TABLE 3. *Characteristics of lake water of Fig. 15*

	Number of lakes	P total (μg/L)	N total (mg/L)	Chloro A (μg/L)	Z Secchi (m)	DOC (mg/L)	UV254* (a.u.)
Oligotrophic	6	4–12	0.2–0.3	1–2.5	3–7.5	2.5–5	0.06–0.15
Mesotrophic	9	10–23.5	0.31–0.52	3–7.5	1.8–3	5.9–15	0.13–0.60
Eutrophic	6	21.5–100	0.7–1.3	11.5–30	0.7–1.5	11–22	0.35–0.86

*For a pathlength of 10 mm at 254 nm.

different from one lake to another. The main explanation is related to the nature and content of natural organic matter, but also to the trophic state of the water body.

The trophic state is deduced from the measurement of some physico-chemical parameters: the concentration of total phosphorous (and nitrogen), chlorophyll A and the measurement of the transparency of water with the Secchi disk. The trophic classes are given from the raw values [20] or after calculation of the trophic state index (TSI) [21]. The three main classes of trophic state (oligotrophic, mesotrophic and eutrophic) correspond to groups of spectra rather different in importance, even if a few of them can be considered with close shapes, for the intermediate class. The typology is confirmed by the study of physico-chemical parameters in Table 3. The values correspond to the usual limits of the trophic states definition for the concentrations of total phosphorous (and nitrogen), chlorophyll A and the measurement of transparency. Moreover, the results of dissolved organic carbon (DOC) and of absorbance at 254 nm (UV254) show a rather good discrimination of the three classes, even if the limits of the intermediate class seem to be fuzzy. The reason why there is a relation between these two approaches is probably linked to the significance of the results. On the one hand, the classical procedure, based on the measurement of the previous parameters, gives information on the conditions and/or the presence of phytoplankton. On the other hand, the use of UV spectrophotometry gives qualitative information (from the spectra shape or UV254 values), as well as quantitative estimation of DOC and nitrate. These complementary parameters of water composition are linked to the conditions and/or presence of microorganisms and plankton. Relatively high concentrations of DOC and nitrate give rise to the increase of biomass in the water body.

The results of this first study on the use of UV spectrophotometry for the characterisation of the trophic state of lakes must be confirmed with other experiments.

A last example concerns the study of a lake and its tributaries. In summer 2004, five sampling campaigns were carried out for the preliminary study of Lake Brome watershed in Southern Québec, Canada. Figure 16 shows the sampling stations of the lake and its tributaries. Four sampling stations were located on the Lake (2–5), one at the outlet (1) and eight on the main tributaries (6–13). Some physico-chemical parameters were acquired, as well as UV spectra of samples. Table 4 presents the synthesis of the results. The main point is that the water quality of the Lake is more stable during summer for temperature, conductivity and pH than the one of tributaries. Obviously, temperature and pH values are higher in the Lake. On the contrary, except for three tributaries, the concentration of dissolved organic carbon is more variable in the Lake, even if the concentrations are lower. The characteristics of Lake Brome, with an area of 14.5 km², a residence time of about 2 years and a maximal depth of 13 m, explain these observations. The set of spectra corresponding to the last campaign of July (Fig. 17) shows a great difference in shape, with spectra characterised by high absorbance values for tributaries (samples 6, 7, 12 and 13), and by low ones for water of the Lake (samples 2–5). The spectrum of

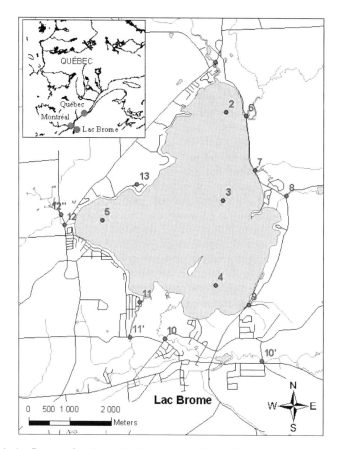

FIGURE 16. *Lake Brome, Southern Québec: sampling stations.*

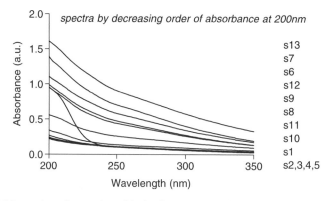

FIGURE 17. *UV spectra of samples of Lake Brome and tributaries.*

TABLE 4. *Characteristics of water and tributaries of Lake Brome (N total and P total have been analyzed once)*

	Temp. (C)		Cond. (μS/cm)		pH (u. pH)		DOC (mg/l)		N total	P total
Sample	Mean	C.V. (%)	Mean	C.V. (%)	Mean	C.V. (%)	Mean	C.V. (%)	(mg/L)	(μg/L)
1	21.8	6.2	116	12.5	7.3	5.8	5.9	25.4	0.22	4
2	21.5	9.1	112	5.4	7.5	1.3	6.0	27.8	0.28	4
3	21.3	6.6	110	6.0	7.5	1.8	5.6	27.8	0.25	3
4	21.3	8.5	110	5.1	7.6	1.8	5.6	28.6	0.29	5
5	21.1	6.9	111	7.6	7.6	1.4	5.7	28.7	0.30	3
6	16.9	2.5	128	20.2	7.2	2.4	17.0	3.2	0.49	15
7	19.2	11.6	71	26.3	7.0	2.7	15.3	11.3	0.47	22
8	18.3	12.3	64	16.5	7.2	3.8	7.5	11.3	0.41	7
9	16.2	14.1	212	25.4	7.2	2.8	7.6	31.3	1.18	4
10	18.0	13.4	110	28.7	7.2	2.9	5.2	30.2	0.31	3
11	19.6	16.8	147	12.5	7.1	1.7	10.9	42.1	0.59	13
12	17.4	18.7	124	22.2	7.1	2.6	14.3	16.1	0.56	14
13	15.9	16.0	220	33.4	7.1	2.0	23.7	0.8	0.46	22

one sample (9) is very different from the others, because of a higher concentration in nitrate (see Table 4).

In contrast to the other tributaries characterised by the presence of natural organic matter, more often coloured, the tributary corresponding to sample 9 is a small one, coming from a golf course and thus probably leaching the excess of nutrients used for greens.

Despite the low values of total phosphorus (analyzed once), the trophic status of Lake Brome must be considered mesotrophic, taking into account the average measurement of transparency (around 2.5 m) and the concentration of DOC.

Figure 18 presents the variation of UV spectra with time for the lake and some tributaries. The maximum of absorbance was measured at the beginning of September for the Lake, contrary to the tributaries where the maximum were in August. This can be explained by a beginning of algae bloom at the end of August, when the total nitrogen concentration becomes limiting, followed by a rainy period. The biomass production was still high in Lake during the September campaign.

Another study is envisaged, for a better monitoring of algae bloom during the end of August, and to point out some diffuse pollution sources, downstream tributaries.

Before leaving this section, one question could be asked about UV-visible characterisation of pigments. Obviously, these compounds are absorbing substances, but taking into account their potential concentration in water, spectrophotometry seems to be limited for the purpose. However, an application for chlorophyll is presented in the next chapter.

3.5. Groundwater study

In contrast to the previous applications, a groundwater quality study is always more difficult because of sampling. In general, a sample must be obviously representative of the characteristics of the studied water body. For surface water, a sampling program can include several samples depending on spatial or time water quality variations. For groundwater, often pumped through wells or boreholes, the local water quality around

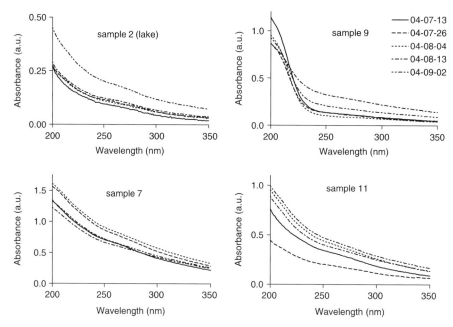

FIGURE 18. *UV spectra of samples of Lake Brome and tributaries: variation with time.*

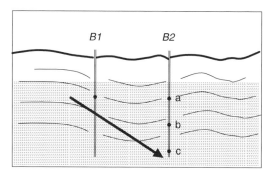

FIGURE 19. *Sampling location for two boreholes.*

the pumping point may be different from the one of the aquifer. Moreover, the material can bring some interference in water quality as, for example, metallic compounds. In order to be sure of the sample quality, it is recommended to pump before sampling during a given time (for example, at least 1–3 h). Considering that the monitoring of water quality is preferable to an arbitrary pumping time, a procedure based on UV spectra evolution has been chosen [22]. The sampling operation is carried out when the UV spectrum shape no longer varies. The conductivity measurement can also be followed during pumping.

A first experiment on groundwater quality study has been conducted in a karstic site in the South of France. Physico-chemical parameters and UV spectra have been acquired for the general study of the aquifer [23]. Figures 19 and 20 show the location and UV spectra of samples from two close boreholes, at different heights, for one of the wells.

The distance between the two boreholes B1 and B2 is 5 m, and their depth is 60 m. The water table was at a depth of 43.5 m below the topographic surface. The area is under the influence of a small river where treated wastewater of a small village (1500 inhabitants) is discharged. The sampling height in B1 was 45 m, while three heights of sampling have been considered for B2 (a 45 m, b 49 m and c 58 m).

Figure 20 clearly shows the existence of a relationship between the two boreholes, B1 near the water table and B2c near the bottom. Moreover, water pumped from B2 was characterised by a higher concentration in organic matter, probably due to the drainage of the above-mentioned river.

Another experiment has been made near Rouen in the North of France, in an alluvial plain bordered by chalk cliffs. Four sampling points have been chosen in order to study the water quality relation between the cliffs and the aquifer of the plain (Fig. 21). Figure 22 shows that the water quality is close for the different samples with an increase in nitrate concentration during its infiltration into the karst (stations 2, 3, 4). The presence of fine particles in sample 4 (borehole) suggests the drainage of the alluvial deposits. Table 5 presents some chemical characteristics of the samples. The quality of the source cannot be explained by the doline composition but only by the karstic drainage. The influence of

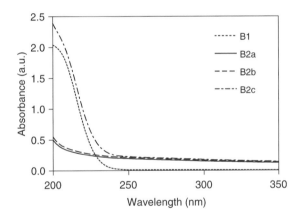

FIGURE 20. *UV spectra of groundwater samples (from Fig. 19).*

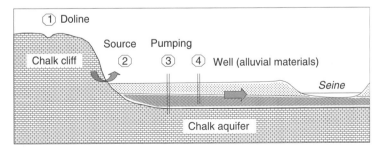

FIGURE 21. *Sampling site near Rouen.*

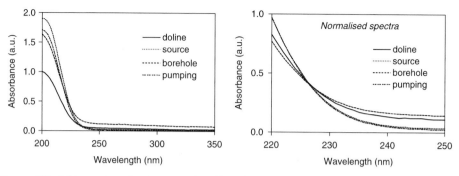

FIGURE 22. *UV spectra of water samples (from Fig. 21).*

TABLE 5. *Physico-chemical characteristics of samples (Fig. 21)*

	Doline	Borehole	Source	Pumping
DOC (mg.L^{-1})	< 1	3	< 1	2.5
NO$_3^-$ (mg.L^{-1})*	13	21	30	26
TSS (mg.L^{-1})*	< 5	29	< 5	< 5

*Estimated by UV.

the doline water composition on the source quality is visible only during wet weather conditions because of important infiltrations [24]. Normalised spectra present an isosbestic point that confirms the quality conservation of water.

4. STUDY OF WASTEWATER DISCHARGE

This part is related to studies of wastewater discharges, generally treated, into rivers and the sea.

4.1. Discharge in river

One of the main applications of UV spectrophotometry is the study of polluted water, in case of wastewater discharge, for example. A first experiment presented in Fig. 23 concerns a local study close to the discharge of treated wastewater of a 150,000-inhabitant biological treatment plant (Aix en Provence, Southern France) into a river with a dilution factor of about 10. The sampling of river water has been carried out at 10 m upstream and downstream of the discharge. UV spectra of river water compared to that of the effluent show a direct isosbestic point (IP), witness of quantity and quality conservation of the main groups of compounds, nitrate and anthropogenic organic matter (see Chapter 2).

The existence of a direct isosbestic point is not always observed in case of the study of wastewater discharge. Another experiment has been realised for the study of the evolution of the Calavon river quality (Southern France), between Apt and Cavaillon (25 km). This part of the river receives the discharge of a 20,000-inhabitant biological treatment plant, with a small dilution factor (5) in summertime.

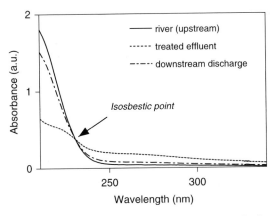

F<small>IGURE</small> 23. *UV spectra of water samples around a wastewater discharge.*

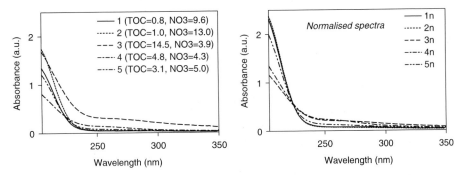

F<small>IGURE</small> 24. *UV spectra (raw and normalised) of the Calavon River (concentration in mg/L).*

Figure 24 shows the evolution of UV spectra, between the upstream (sampling points 1 and 2) and downstream (sampling points 4 and 5) areas of the discharge (sample 3). From the discharge point, the distances are respectively −3 km, −100 m, +100 m and +15 km. The treatment efficiency of the treatment plant is quite good, as the pollution parameter values are low (TOC = 14.5 mg/L, COD = 67 mg/L, BOD5 = 13 mg/L).

Considering the shape of UV spectra of the river water and treated wastewater, a hidden isosbestic point (HIP) can only be found after normalisation. In this case, the spectra set evolution clearly shows a dilution of anthropogenic matter discharged (and the partly self purification of river water) and of nitrate concentration of the upstream river water. The presence of an HIP is related to the quality conservation of water (see Chapter 2).

Another study based on the use of UV spectrophotometry for the evaluation of the impact of treated wastewater discharge in rivers [25] has shown that not only the qualitative and quantitative evolution of river water quality was possible, but also some hydraulic parameters such as the dilution factor of discharge or confluences.

The dilution factor (F) of a mixture can generally be calculated as follows:

$$F = \frac{Qm}{Qd} = \frac{Cd - Ci}{Cm - Ci}$$

where Qd and Qm are the flow rates of the treatment plant discharge and of the mixture downstream of the discharge, respectively, and Cm, Cd and Ci are the respective concentrations of any conservative parameter in the mixture, in the discharge and in the river upstream of the discharge.

The calculation of dilution factors has been validated in this same study [25] from data of TOC, nitrate and anionic surfactants, either measured by reference methods or UV-estimated.

4.2. Discharge in sea

In seawater, the spatial impact of treated wastewater discharge is limited, considering the high dilution potential of the sea. Two experiments have been carried out on the Mediterranean Sea, near Marseille in South of France, with UV spectra acquisition.

The first concerned the study of treated wastewater discharge of the Marseille urban area (2,000,000 inhabitants). Figure 25 presents the evolution of UV spectra of seawater, sampled (at the surface of the sea, from a boat) from the discharge, along the direction of the main current.

The influence of the discharge is visible up to 30 m, mainly because of the presence of relatively high residual organic matter in treated wastewater (COD around 100 mg/L), due to the physico-chemical process of the treatment plant. The TSS concentration (around 30 mg/L) contributes to the turbidity effect in the mixture. The treatment plant extension that should include a biological process would lead to the reduction of wastewater impact in this sea area.

The second experiment has been conducted not far from the previous one, on a petrochemical site (Naphtachimie, Lavéra), integrating one of the largest cracking unit of

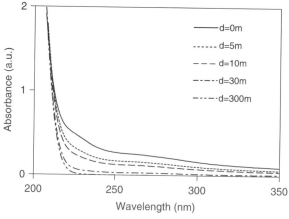

FIGURE 25. Dilution of treated wastewater of the Marseille area, discharged in the Mediterranean Sea.

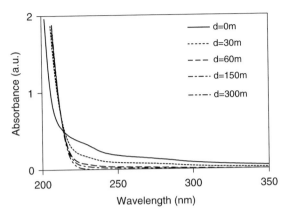

FIGURE 26. *Dilution of treated wastewater of a petrochemical site, discharged in the Mediterranean Sea.*

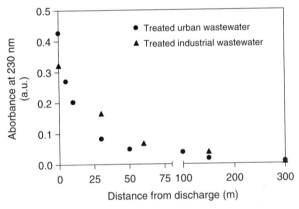

FIGURE 27. *Evolution of absorbance at 270 nm with the distance from two discharges.*

the world, producing almost 1,000,000 tons of ethylene a year. A very efficient biological wastewater treatment plant treats all wastewater of the site plus some external wastes (see Chapter 10). Several samples were taken from a boat in the discharge area [26]. As for the last study, spectra present a high absorption below 220 nm due to the presence of high concentration of chloride in seawater (Fig. 26). Figure 27 shows the evolution of the absorbance value at 270 nm with distance for the two examples.

The organic and particle load of the treated wastewater discharged in sea is lower than the one of Marseille urban area. But the dilution of wastewater remains always visible at a distance of about 100 m. Actually, in contrast to the previous case, the discharge site is not located in an open sea, but in a relatively small bay.

4.3. Accidental discharge

Before studying drinking water quality, two last applications are presented for the diagnosis of accidental pollution in rivers.

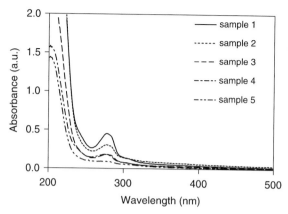

FIGURE 28. *UV-visible spectra acquired during accidental discharge into a river.*

UV spectra of several samples collected inside one industrial site (samples 1, 2) and in the polluted river (samples 3, 4, 5) have been acquired (Fig. 28) further to the death of a lot of fishes (for the sake of confidentiality, no precise indication will be given). The presence of organic pollutants of phenolic nature has been suspected from the specific signal at 278 nm (Chapters 3 and 10). This hypothesis has been confirmed by a bathochromic shift in basic medium. It can be observed that a dilution factor occurs for samples 3 and 4. Sample 5 has been taken two days after the incident, and it can be noticed that the pollutant concentration has clearly diminished. Another observation can be made related to the high absorbance values below 250 nm, probably associated to a high mineralisation of water. The conductivity measurement of samples has confirmed this fact with values between 3 and 8 mS.cm^{-1}, the higher values corresponding to industrial samples. This real experiment has been followed by the diagnosis of the pollution origin.

A last example is given in Fig. 29. The spectrum of polluted river water shows a shape close to the one of a diluted raw wastewater with a significant concentration of surfactants

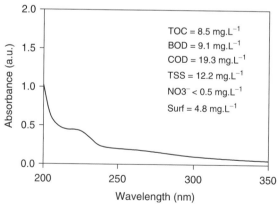

FIGURE 29. *Surface water spectrum containing a mixture of anthropogenic and natural organic matter with an important concentration of surfactants.*

resulting in the existence of an important shoulder near 220–230 nm (due to the aromatic ring; see Chapters 3 and 4).

Surfactants are usually an interesting fingerprint of domestic pollution, responsible for the relatively high values of aggregate parameters (TOC, BOD, COD). In this case, a simpler visual characteristic can be used for pollution diagnosis, the presence of white persistent foam, particularly in aerated area, related to the presence of surfactants.

Even if its interest has been demonstrated for these applications, UV spectrophotometry cannot show the presence of nonabsorbing species or of absorbing compounds the concentration of which is too low with regard to their optical properties (absorptivity values).

5. DRINKING WATER QUALITY

5.1. Mineral water quality

There exists a lot of mineral and source water for drinking. In some countries, it is mandatory to display water composition on the bottle label. Spectra of several mineral and source water have been acquired (Fig. 30), three among the most known French mineral water brands (Volvic, Vittel and Evian) and two French source water brands from Corsica (Zilia and Saint Georges).

The different spectra have the same shape due to the very good quality of water (drinking water contains a negligible concentration of organic matter and is free of microorganisms) and to a variable content of nitrate from one mineral water brand to another.

Table 6 presents the concentration of major mineral compounds obtained from the respective labels. The selected waters represent a large variety of mineral composition from a carbonato calcic and magnesian matrix (Evian) to a sulphato-carbonato calcic one (Vittel), with a few mineralised waters (e.g. Volvic). Notice that the chosen source waters are also characterised by a low mineralisation and a relatively high concentration of chloride with regard to mineral waters.

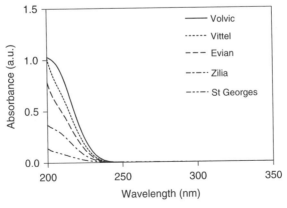

FIGURE 30. *UV spectra of French mineral and source water brands.*

TABLE 6. *Composition of some French mineral and source water brands*

Concentration (mg.L^{-1})	Evian	Vittel	Volvic	Saint georges	Zilia
Ca^{2+}	78	202	9.9	5.2	11
Mg^{2+}	24	36	6.1	2.4	5.1
Na$^+$	5	3.8	9.4	—	15
K$^+$	1	—	5.7	1.2	1.3
HCO$_3^-$	357	402	65.3	30.5	67.7
SO$_4^{2-}$	10	306	6.9	—	5
Cl$^-$	4.5	—	8.4	25	15
NO$_3^{-a}$	3.8	—	6.3	—	2.2
NO$_3^{-b}$	4.7	6.3	7.9	<0.5	2.3

^{a}Label value.
^{b}UV measurement.

The last point is that, except Saint Georges, all the other mineral water brands have significant concentration of nitrate, much lower than the European regulated value for drinking water, for example, but not negligible. The nitrate concentration in water is always slightly increasing, even in mineral and source waters, and this phenomenon will probably continue for a few years despite the collective efforts for N treatment.

5.2. Production of tap water

The last application in this chapter deals with the study of water quality evolution in a treatment plant for tap water production. The plant is located close to Montpellier in the

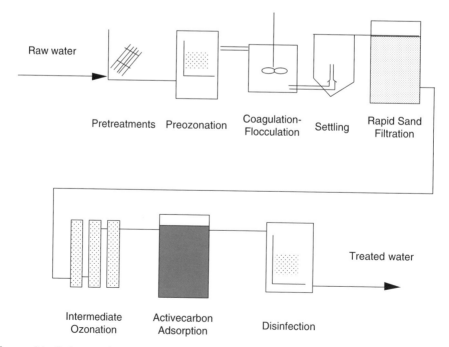

FIGURE 31. *Scheme of water treatment plant.*

South of France and supply tap water for almost 100,000 inhabitants in a very touristy area close to the sea. This plant is composed of all efficient processes for water production from surface water and groundwater pumping (Fig. 31). Figures 32 and 33 show the evolution of UV spectra during treatment.

After a pretreatment step, followed by a preozonation, a physico-chemical process (coagulation–flocculation, settling and sand filtration) removes the suspended solids and colloids of water. Then the dissolved organic compounds are oxidised and adsorbed with their by-products on active carbon. Before leaving the plant, a last oxidation step allows the disinfection of the produced water and to assure its sanitary protection along the distribution network, until its consumption.

The resource is composed of two main types of water, one coming from a canal and the second from a regional aquifer. Figure 26 shows that surface water is mainly characterised by some organic matter while groundwater contains a relatively high concentration of nitrate (about 15 mg/L). The quality of treated water is also characterised by the presence of nitrate (12.5 mg/L) and a very low content of organic matter (<1 mg/L). Figure 29 displays the effect of each treatment step on water quality, with a general decrease in

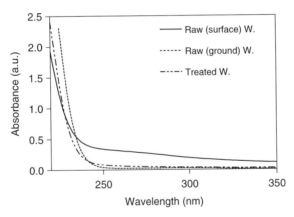

FIGURE 32. *Effect of water treatment on UV spectra (pathlength 50 mm).*

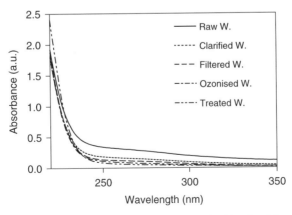

FIGURE 33. *Evolution of UV spectra along tap water production (pathlength 50 mm).*

organic matter and in nitrate concentration due to the income of groundwater during the treatment (at the entry of the coagulation–flocculation basin).

References

1. N. Ogura, T. Hanya, *Nature*, 212 (1966) 758.
2. R.A Dobbs, R.H. Wise, R.B. Dean, *Water Res.*, 6 (1972) 1173.
3. P. Forster, *Water Res.*, 19 (1985) 701.
4. R. Briggs, K.V. Melbourne, *Proc. Soc. Wat. Treat. Exam.*, 17 (1968) 107.
5. M. Mrkva, *Water Res.*, 9 (1975) 587.
6. M. Mrkva, *Water Res.*, 17 (1983) 231.
7. O. Thomas, F. Théraulaz, M. Domeizel, C. Massiani, *Environ. Technol.*, 14 (1993) 1187.
8. O. Thomas, F. Théraulaz, C. Agnel, S. Suryani, *Envir. Technol.*, 17 (1996) 251.
9. O. Thomas, F. Théraulaz, V. Cerdà, D. Constant, P. Quevauviller, *Trends Anal. Chem.*, 16 (1997) 419.
10. R.C. Hoather, *Wat. Treatm. Exam.*, 2 (1952) 9.
11. O. Thomas, S. Gallot, N. Mazas, Fres. *J. Anal. Chem.*, 338 (1990) 238.
12. H. De Haan, *Freshwater Biol.*, 2 (1972) 235.
13. J. Buffle, Complexation Reactions in Aquatic Systems: an Analytical Approach. Ellis Horwood Limited (Ed), New York, (1988).
14. O. Thomas, S. Gallot, E. Naffrechoux, *Fres. J. Anal. Chem.*, 338 (1990) 241.
15. H. El Khorassani, H.G. Besson, O. Thomas, *Quimica Analytica*, 16 (1997) 239.
16. W.J. Maier, *Multiwavelength Absorbance Measurements for Monitoring Trace Organics in Water, in Chemistry in Water Reuse*, Vol. 1, W.J. Cooper Ed., Ann Arbor Science, (1981).
17. R. Sutton, G. Sposito, *Environ. Sci. Technol.*, 39 (2005) 9009.
18. S. Vaillant, *Organic Matter of Urban Wastewater: Characterisation and Evolution.* PhD thesis, Université de Pau et des Pays de l'Adour, France (2000).
19. V. Mesnage, S. Bonneville, B. Laignel, D. Lefebrre, J.-P. Dupont, D. Mikes, *Hydrobiol.*, 475/476(2002) 423.
20. R.A. Vollenweider, J.J. Kerekes, *Synthesis Report, Cooperative Programme Monitoring of Inland Waters*, OCDE Ed., Paris, (1980).
21. R.E. Carlson, *Limnol. Oceanogr.*, 22 (1977) 361.
22. O. Thomas, F. Théraulaz, *Trends Anal. Chem.*, 13 (1994) 344.
23. T. Winiarski, O. Thomas, C. Charrier, *J. Contam. Hydro.*, 19 (1995) 307.
24. N. Massei, M. Lacroix, H.Q. Wang, B. Semega, J-P. Dupont, *Transport of Suspensed Solids Through Saturated Porous Media: Experimental Approach and Field Measurements.* In Proc. AIH Conference, Le Cap, Balkema Ed., Amsterdam, 2000.
25. H. El Khorassani, F. Théraulaz, O. Thomas, *Acta Hydrochim. Hydrobiol.*, 26 (1998) 296.
26. H. El Khorassani, *Characterization of Industrial Effluents by UV Spectrophotometry: Application to Petrochemical Industry.* PhD thesis, University of Aix-Marseille I, (1998).

CHAPTER 8

Urban Wastewater

O. Thomas[a], F. Theraulaz[b], S. Vaillant[c], M.-F. Pouet[a]

[a]Observatoire de l'Environnement et du Développement Durable, Université de Sherbrooke, Sherbrooke, Québec, J1K 2R1, Canada; [b]Laboratoire Chimie et Environnement, Université de Provence, 3 place V. Hugo, 13331 Marseille Cedex, France; [c]Alstom Power, Bâtiment 86, 3 avenue des 3 chênes, 90000 Belfort, France

1. INTRODUCTION

Urban wastewater is most likely the best experimental field for the application of UV spectrophotometry. The first reason is that organic pollution is composed of unsaturated compounds, which enables finding, for any sample, simple relations between its UV spectrum and the value of aggregate parameters such as COD, BOD, etc. (see Chapter 4). As a consequence, many researches have been conducted on this topic for more than 30 years [1–3]. All these authors have proposed the measurement of the absorbance at 254 nm. Even if this wavelength seems to correspond to a shoulder in the spectrum, the reason for the choice of this wavelength was more practical (254 nm being the main radiation below 300 nm of low-pressure mercury lamp), than scientific, because of possible interferences from industrial discharges.

But even if more adapted methods for spectra exploitation are nowadays used for the estimation of some water quality parameters, the main interest of using UV spectrophotometry for urban wastewater quality monitoring is the significance of UV spectrum shape, the evolution of which gives useful information.

First, this chapter presents some studies concerning the evolution of raw wastewater quality in sewers and, second, the effect of some treatment processes. Two examples of treatment plants are then presented, before a synthesis leading to the proposal of a wastewater classification.

2. SEWERS

2.1. Fresh domestic effluent

Raw urban wastewater is composed of fresh domestic effluent, the quantity and quality of which vary according to time and space (Fig. 1). Two other factors must be considered in order to explain some specific variations: rainfall for unitary sewers, and industrial discharges for all types of sewers.

The evolution of a wastewater UV spectrum is always the same from the source to the discharge after treatment. The shape is always decreasing except for short wavelengths where nitrate formation can lead to high absorption. Another observation to be made

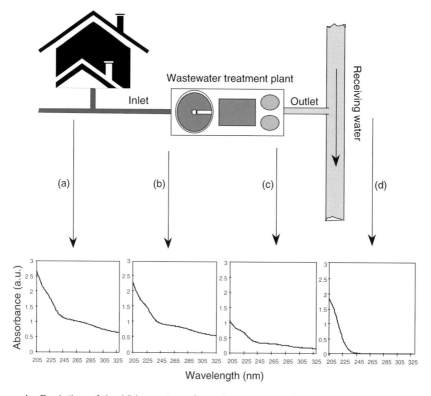

is that UV spectra of domestic or urban wastewater have practically always the same shape, whatever the country.

A fresh domestic wastewater is composed of soluble, colloidal and suspended constituents produced by the physiological and domestic activity of human beings. Urine, faeces, as well as detergents, particulate matter, grease and several other common pollutants are daily rejected with wastewater. Among all these compounds, urine is probably the most important with respect to UV absorption and organic (and nitrogen) pollution concentration. Detergents also strongly absorb (particularly the benzenic forms) and can explain the peak often encountered at 225 nm (see Chapter 4). Figure 2 shows the UV spectra of human urine and an anionic surfactant, both of which can be considered as the main factors explaining the shape of the UV spectrum of fresh domestic wastewater.

As the simpler evolution factor of wastewater is certainly due to the presence of particles, it is interesting to study the coarse distribution of their size. A simple experiment can be carried out on a wastewater sample [4]. The UV spectrum of the raw sample is compared to the spectra of filtrates obtained after filtration 1 μm, 0.45 μm, 0.1 μm, 0.01 μm (Fig. 3). The effect of the different fractions constituting the colloidal and solid phases can be related to spectra variation, the absorbance of which decreases as the filtration step advances.

Figure 4 shows that the different spectra corresponding to the various cutting sizes can either be featureless or present a specific shape. In the latter case, the adsorption

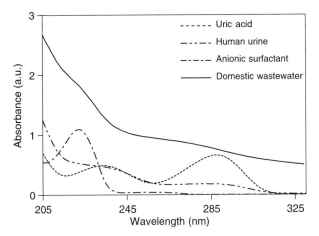

FIGURE 2. *UV spectra of urine (dilution 400), DBS 10 mg/L (dodecylbenzenesulfonate) and fresh domestic wastewater (dilution 2).*

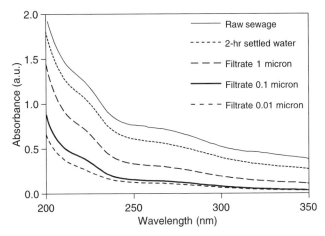

FIGURE 3. *UV spectra of urban wastewater before and after several filtration steps.*

of some organic compounds such as benzenic surfactants may be responsible for the shoulder at 225 nm on the difference spectrum corresponding to smaller colloids (between 0.01 and 0.1 μm). Another observation already made in Chapter 6 is that the average slope of spectra increases as the particles size decreases.

2.2. Variation of quality according to time

Municipal wastewater is obviously related to the water supply consumption of inhabitants and of municipal and security services (watering, washing). Wastewater flow varies according to the season of the year, weather conditions, day of the week, and time of

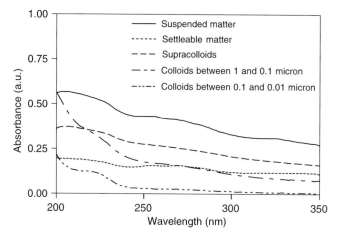

F IGURE 4. *Differential spectra of urban wastewater before and after several filtration steps (from Fig. 3).*

the day. Under dry weather conditions, daily wastewater flow shows a diurnal pattern close to the water demand.

Wastewater flow fluctuations depend on the sewers' storage volume and on the time required for the wastewater to reach the treatment plant. Industrial discharges tend to reduce the peak flows. In small communities, two daily peaks are generally observed, while only one is noted in larger cities. In the latter case, the length and complexity of the sewer network tend to smooth the daily flow variation [5].

The variations of pollution fluxes or load (BOD_5, quantity for instance) depend on flow variations. Thus, pollution quantity as well as the effluent quality may vary, on one hand, according to the network size and the importance of industrial discharges and, on the other hand, according to the type of sewer, separated or combined (i.e. also collecting rain water).

Figure 5 presents the evolution of some UV spectra of raw sewage sampled hourly during 24 h at the entrance of a wastewater treatment plant (after the grit chamber) [4].

In this example concerning a medium urban area of about 10,000 inhabitants, the sewer system is mainly of separated type. The corresponding values of the main parameters are also reported in Fig. 5, so that the spectra correspond to the filtered samples and to suspended solids.

The study of these UV spectra enables the survey of the evolution of both the pollution quantity (by the estimation of parameters) and the quality of the effluent.

The relation between concentration and flow is clearly shown in this example. Indeed, flow rates and UV spectra present the same variation. During the night, the sample presents a spectrum characterised by a monotonous shape without any shoulder. This can be easily related to a low domestic activity. On the contrary, day time spectra present a more important pattern, with the shoulder associated to the presence of surfactants. A correlation between the values of flow rates and aggregate parameters plus some specific compounds (detergents) could easily be established. This example shows that the knowledge of time variation in wastewater quality is very important for sewer and treatment plant management.

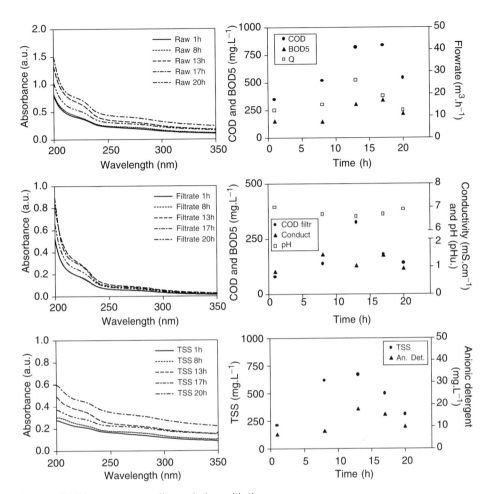

FIGURE 5. *Wastewater quality variation with time.*

2.3. Evolution along the sewer

The variation of wastewater quality is related to the human activities (and thus time dependent as shown before) but also to some physico-chemical factors occurring along the sewer. Fresh organic matter composition can vary from the source (houses, for example) to the treatment plant because of physico-chemical phenomena (solids settling, phase transfer, oxidation, etc.).

Figure 6 shows two sets of spectra of urban wastewater sampled at the beginning of a sewer network (upstream point) and at the inlet of the treatment plant [4]. Sampling has been made at the beginning of the afternoon, taking into account the transfer time from one station to another. Some differences have to be noticed between spectra. Firstly, the spectrum of wastewater sampled upstream of the network is more important than the one sampled at the inlet of the wastewater treatment plant. The corresponding values of the main parameters explain this observation with, respectively, 1055 and 818 mg/L

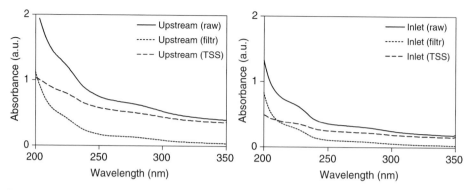

FIGURE 6. *Wastewater evolution along a sewer. UV spectra (dilution 5) of raw and filtered samples and spectra corresponding to TSS.*

for COD and 467 and 315 mg/L for BOD$_5$, from upstream to the treatment plant. This variation in concentrations can be due to several reasons such as the input of cooling, watering or even clear parasite water into sewer, or the partial biodegradation of organic matter.

The comparison of differential spectra between the raw and the filtered sample is interesting (Fig. 6). These spectra are related to the characteristics of suspended solids of samples. The TSS concentrations are, respectively, 502 and 669 mg/L, from upstream to the treatment plant, meanwhile the corresponding spectrum seem to be divided twice. The observed increase can be explained, on one hand, by solids input (by incoming water) or formation (related to the biodegradation), and on the other hand, by the aggregation of colloids in suspended solids [6]. This is confirmed by the fact that particles of larger size absorb less in the UV region than smaller ones (see Chapter 6).

2.4. Effect of rain

The first striking feature of urban storm runoffs is their great variability, with quite different pollution values (loads and concentrations) from site to site [7]. This variability is closely linked to the nature of storm events. Indeed, each rainfall is different from the other and induces specific phenomena [8]. Nevertheless, average values still stand out, which enables a comparison with wastewater in dry weather conditions (Table 1) [9].

TABLE 1. *Average concentration for combined sewage and wastewater [9]*

Parameter	Urban storm runoff	Wastewater
TSS (mg/L)	176–2500	150–500
Volatile fraction of TSS (%)	40–65	70–80
Chemical oxygen demand (mg O$_2$/L)	42–900	300–1000
Biological oxygen demand (mg O$_2$/L)	15–301	100–400
COD/BOD$_5$ ratio	3.4–6.0	2
Nitrogen total Kjeldhal (mg N/L)	21.0–28.5	30–100

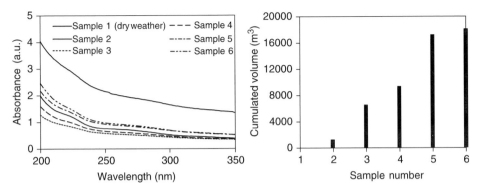

FIGURE 7. *Evolution of wastewater UV spectra during a storm event (all samples were diluted twice and thus absorbance values multiplied twice).*

This comparison shows a dilution effect for dissolved pollution (pointed out by NTK) but an increase of suspended solids. Another conclusion is a change in the nature of the suspended pollution, mostly organic in dry periods and more mineral during rains. This phenomenon is mainly explained by the suspended matter influx, leading to a concentration maximum that is five times more important than for wastewater. Moreover, the organic load seems to be less biodegradable, as pointed out by a higher COD/BOD$_5$ ratio.

Figure 7 presents the evolution of urban sewage during a storm event [10]. The dry weather sample was taken after the grit chamber and the wet weather one at the overflow of the main interceptor leading to the treatment plant. The sewer system in this town is combined in the centre and separate in the outskirts. The first wet weather sample (sample 2) is highly concentrated since it shows absorbance values twice as high as the ones of dry weather (sample 1). After this first load, the following sample (sample 3) is diluted showing the lowest absorbance values of the spectra set. Then, sample concentration slowly increases to become slightly higher than during dry weather (sample 1) at the end of the rain (sample 6).

Normalised spectra are compared in Fig. 8, showing that the shape of urban storm runoff spectra is rather diffuse at the beginning of the rain. During the storm event, the slope between 200 and 240 nm gradually becomes steeper while the absorbance above 250 nm decreases relatively. Spectra shape thus tends to progressively look like the one of dry weather spectrum as the rain gets closer to the end. This phenomenon takes place around an isosbestic point located at approximately 234 nm, showing that sewage can be compared to a mixture of two major components [11]. The steeper slope in the first part of dry weather spectra is an indication that soluble matter is preponderant during dry weather conditions, while the high absorbance values above 250 nm suggests that suspended solids predominate during rains.

Sample analysis (Table 2) confirms the information brought by UV spectra observation showing that the first wet weather sample (sample 2) is heavily loaded in terms of COD. The following is diluted (sample 3) with a lower COD and conductivity, but with a higher SS concentration than during dry weather (sample 1). After the first part of the rain (samples 4, 5, 6), COD values are slightly higher than those corresponding to dry weather.

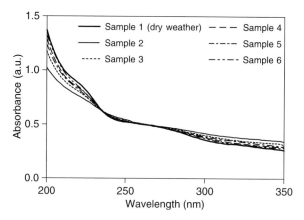

FIGURE 8. *Evolution of wastewater normalised UV spectra during a storm event.*

TABLE 2. *Evolution of nonspecific parameters during a wet weather period and comparison with wastewater in dry weather conditions [12]*

		1	2	3	4	5	6
Weather		Dry	Rainy	Rainy	Rainy	Rainy	Rainy
Conductivity	mS/cm	0.78	0.48	0.38	0.66	0.65	0.86
COD	mgO$_2$/L	430	2400	310	560	590	510
Raw	%	29	90	90	75	61	50
Settleable	%	31	7	4	19	25	33
Supracolloidal	%	40	3	6	6	14	17
Soluble							
TSS	mg/L	207	1489	437	319	375	248
Raw	%	54	87	85	79	77	56
Settleable	%	46	13	15	21	23	44
Supracolloidal	%	88	63	66	79	81	86
TVS*							
DOC	mg/l	53	25	15	25	35	38

*Total volatile solids (fraction of TSS of organic nature).

Figure 9 shows the UV spectra of colloidal and particulate fractions of a sample collected during the first storm period (sample 2).

UV spectra of TSS and raw wastewater compared to UV spectra of sample during dry period (see Fig. 4) are featureless because of the great proportion of settleable particles (90% of TSS) that diffuse light by diffraction. Chemical response is very weak: for example, the shoulder, especially at 225 nm, does not appear to show that the physical response dominates (see Chapter 6).

2.5. Synthesis and other applications

Figure 10 presents a synthesis of the main phenomena involved in wastewater quality variation along a sewer network, emphasising on solid transfers during dry and

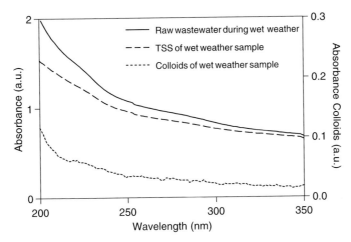

FIGURE 9. *UV spectra and difference spectra of sample 2 of Fig. 7 (diluted twice).*

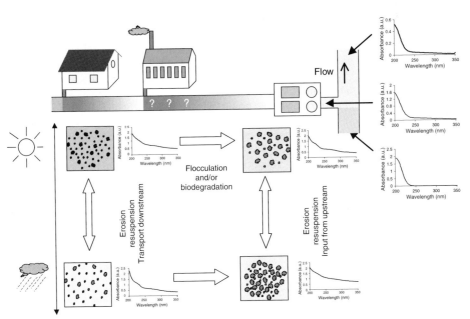

FIGURE 10. *Organic matter of urban wastewater: evolution along sewer [4].*

wet weather. This presentation shows the great interest of using UV spectrophotometry in the understanding of phenomena.

Another application not presented in this chapter is the diagnosis of wastewater quality in sewer for the detection of non-allowed industrial discharge, for example. In case of suspicious junction, it is easy to plan an experiment with sampling for parameters measurement and UV spectra acquisition. The checking of particular wastewater quality will be explained in the next chapter.

Some other use of UV spectrophotometry for urban wastewater characterisation can be equally envisaged. For example, the UV study of gel chromatography fractions leads to the revelation of various groups of compounds that can be classified with their molecular weight. This will be presented in Chapter 10.

3.   TREATMENT PROCESSES

The application of UV spectrophotometry for the study of treatment processes is dependent on the user's needs. From primary treatments to the global efficiency estimation of the treatment plant, several applications are presented in this section.

3.1.   Primary settling assistance

The purpose of primary sedimentation is to remove settleable solids. About 50–70% of TSS and 30–40% of the BOD$_5$ can be eliminated during this operation, which necessitates a residence time in the clarifier of about 2 h. Figure 11 shows the settling effect on UV spectra, and spectra related to settleable matter and suspended matter (after filtration at 1.2 μm) are presented in Fig. 12.

During settling, the decrease in absorbance on the whole spectrum, especially between 230 and 350 nm where TSS are responsible for a diffuse absorbance, indicates the removal of settleable solids. About 54% of the suspended solids and 29% of the COD are removed by settling (Fig. 11). Actually, settleable solids are also constituted of a part of colloids that can be trapped with larger solids during the settling operation. Their corresponding spectrum represents about 35% of the absorbance due to suspended matter (Fig. 12).

In order to help the user in the design of a settling tank (to build or to check), a simple experiment can be proposed (Fig. 13). A wastewater sample is introduced into a 2-m-high glass column of 100-L capacity, and aliquots are carefully withdrawn for different heights and times. UV spectra are acquired for a quick estimation of the TSS (after a previous

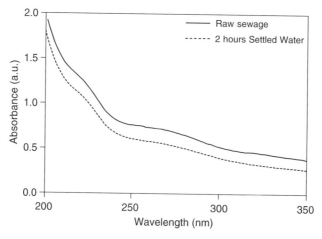

FIGURE 11. *Spectra of raw and 2-h settled sewage.*

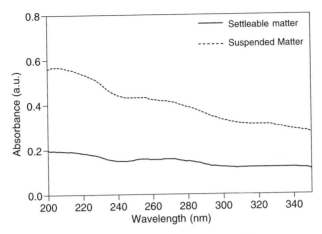

FIGURE 12. *Spectra of the settleable and suspended matter from raw sewage.*

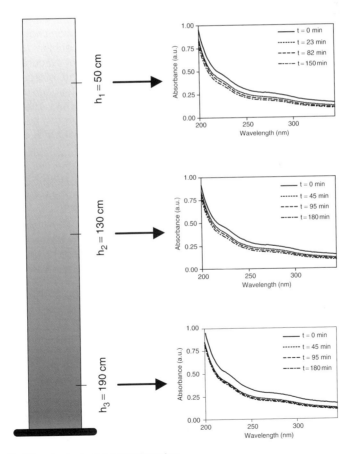

FIGURE 13. *Settling test of urban wastewater.*

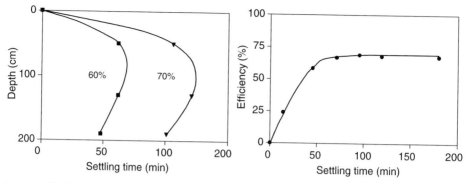

FIGURE 14. *Settling study. Iso-TSS removal yield curves and evolution of settling efficiency with time (for H = 2m).*

calibration if needed). The aim of the experiment is to study the value of the settling yield in function of time or height (Fig. 14).

The settling test is carried out over 3 h. The UV spectra evolution shows that the signal decrease with time is generally quick at the beginning of the experiment (until 30–45 min), and slows down afterwards. These phenomena are typical of hindered settling. This tendency is available for all heights studied (Fig. 13). At the end, all spectra are quickly similar due to the thickening of the settled solids bed. After the test, the settling efficiency can be studied, from the curves of iso-TSS removal yield (Fig. 14). Then, the efficiency evolution with time can be represented for a given height. From this curve, it is easy to determine the settling time and hence the climbing velocity to apply, corresponding to a given efficiency.

3.2. Physico-chemical treatment assistance

The purpose of chemical treatment based on coagulation–flocculation is to eliminate suspended solids and colloids, organic as well as mineral, from wastewater. The desta-bilisation of colloids by the addition of coagulants (such as ferric chloride) leads to the agglomeration of particulate matter. Flocculation can be improved by the addition of organic or mineral polymers (called flocculants). The separation of the chemical sludge produced by coagulation–flocculation from treated water is realised by settling. Such a treatment leads to a yield of 80–90% of TSS and 70–80% of the COD and BOD_5. Apart from phosphorus, soluble matter is not removed.

3.2.1. Jar test

Many variables affect the mechanisms of coagulation–flocculation. The selection and optimum dosage of coagulants and flocculants, the determination of optimum pH and the optimisation of operating conditions (mixing energy and time for rapid and slow mixing) are effectively determined by using a laboratory test called Jar test. It is a valuable tool for the optimisation of existing plant operations, and also for a new design or plant expansion.

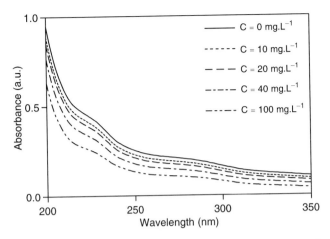

FIGURE 15. *Influence of the amount of FeCl₃ on the UV spectra during a Jar-test.*

This test can be carried out with the help of UV spectrophotometry. Figure 15 presents the evolution of spectra with the coagulant concentration.

The optimum dosage of coagulant corresponds to the lowest residual turbidity or the lowest TSS content in the supernatant. However, the use of turbidity is not always appropriate (see Chapter 6), and TSS measurement is time consuming and therefore too slow to be used for process control. The UV spectrophotometric study enables the control of the effect of ferric chloride dosage. A decrease in absorbance is observed all along the spectrum, especially between 230 and 350 nm where colloids and TSS are responsible for a diffuse absorbance.

The decrease of the shoulder located at around 225 nm, characteristic of anionic surfactants [13], shows that the chemical process enables the removal of a part of detergents. Other dissolved matter that can be mostly noticed between 200 and 240 nm, inducing an absorbance with a steep slope, are not removed.

UV spectrum of urban effluent from a chemical treatment plant has thus a featureless shape, like the one of raw wastewater. The absorbance value globally decreases as the wavelength increases. However, the effect of the coagulation–flocculation process can be noted. Indeed, the absorbance value above 250 nm is very low, showing that the concentration of suspended solids and colloids the optical, effect of which being more sensible in this wavelength range (see Chapter 6), is also very low.

3.2.2. Problem of sample aging

A physico-chemical process based on coagulation–flocculation produces treated wastewater mainly composed of soluble and colloidal matter. A part of supracolloids responsible for TSS is always present, their remaining concentration depending on the coagulant dose. The stability of the treated wastewater is not well known, and it is not rare to see this type of sample aging with an increase in TSS concentration.

For example, Fig. 16 presents the evolution of the TSS concentration of a sample conserved at 4°C. The measured concentrations, as well as the UV estimation based on

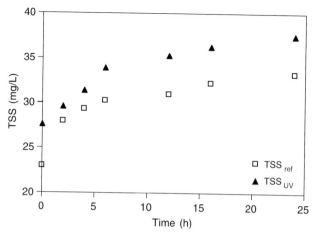

FIGURE 16. *Evolution of TSS concentration versus time (sample conserved at 4°C).*

the use of the coefficient contribution of TSS reference spectra (see Chapter 6), increase with time up to 25–50% of the initial value. Figure 17 shows the normalised spectra and the spectral contribution of TSS and colloids. This experiment can be completed by the study of the evolution with time of contribution coefficients of the different fractions involved in the phenomenon: suspended solids, colloids and surfactants (Fig. 18).

The evolution of UV spectra during aging shows a small decrease of absorbance values for the short wavelengths and a corresponding increase for the higher ones. This is due to both an increase in suspended solids and a decrease in colloids, clearly shown in Fig. 17, resulting in a transfer from the colloidal fraction to the supracolloidal one (leading to a TSS increase). The finest colloids flocculate and produce suspended solids, perhaps aggregated on existing particles. In case of a physico-chemical treated waste-water, the remaining suspension treated by coagulation–flocculation is not stable, and the agglomeration process is not completed.

In order to have a better understanding of the phenomenon, the evolution factors are studied. Process of agglomeration/dispersion can be influenced by the presence of a residual coagulant, but the study of residual iron has shown a weak decrease with

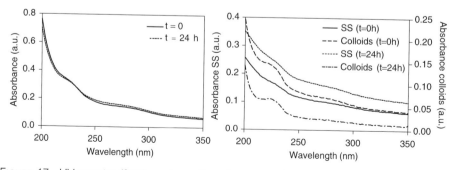

FIGURE 17. *UV spectra (fresh and aged) and spectra of SS and colloids.*

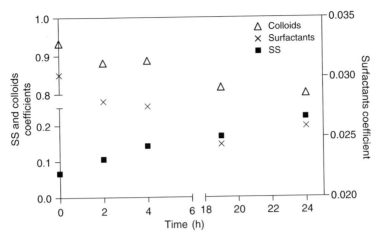

FIGURE 18. *Evolution of TSS and colloid coefficients versus time.*

time and cannot explain the sample evolution. The evolution of surfactants contribution coefficient shows, on the contrary, a decrease in their quantity, particularly in the first part of the sample aging, whatever the coagulant dosage (Fig. 18).

Surfactants have a role of dispersion stabilising almost any suspension. More-over, they can adsorb themselves on solids [14], thus destabilising the suspension. So, with the surfactants' concentration decreasing in solution, the colloids can flocculate or adsorb themselves on suspended solids, explaining both the decrease of colloids and surfactants in the sample.

3.3. Biological processes

Biological processes are widely used for wastewater treatment. Because an important part of organic matter of wastewater is biodegradable, the presence of microorganisms in a process will degrade this main pollution form. The concerned fraction is estimated by the BOD_5 measurement (see Chapter 4), the principle of which is to reproduce this phe-nomenon and to quantify the corresponding oxygen consumption (demand). In contrast to the physico-chemical process, the aim of biological processes is to remove the soluble biodegradable part of organic matter. This is carried out generally in the presence of oxy-gen in an aeration basin and lead to the (partial) mineralisation of matter and to sludge production. This last point is very important because soluble products are converted into solids easily removed by settling in a clarifier (and recirculated).

The interest in UV spectrophotometry for the study of biodegradation has already been demonstrated in Chapter 4 with the study of samples during BOD measurement evolution. The same evolution is more or less observed for biological wastewater treatment plants.

In Fig. 19 are presented the spectra of inlet and outlet of a large treatment plant char-acterised by a rather short mean residence time in the aeration basin (few hours). This process, working with a high-load F/M ratio, does not lead to the complete biodegrada-tion of organic matter as it is shown with the outlet spectrum shape. The efficiency is quantified by removal yields calculated from different aggregate parameters (Table 3).

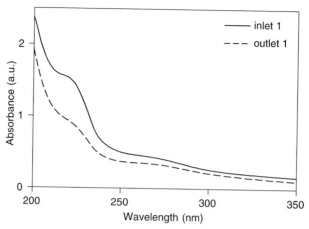

FIGURE 19. *Spectra of inlet and outlet of biological WWTP (high load ratio).*

TABLE 3. *Removal yield for the studied biological wastewater treatment plants*

	WWTP1	WWTP2	WWTP3
Type	High load	Low load	Denitrification
COD inlet/outlet*	695/242	417/85	837/22
Yield%	65	80	97
BOD inlet/outlet*	312/96	179/24	343/<5
Yield%	69	86	>98
TOC inlet/outlet*	44/29	38/7	73/8
Yield%	34	81	89
TSS inlet/outlet*	204/82	146/36	501/10
Yield%	60	75	98
NGL outlet*	58	30	2

*mg/L.

A second small biological wastewater treatment plant characterised by a longer mean residence time in the aeration basin, up to one day or more, has been studied (Fig. 20). In this case, corresponding to a low-load biological process, the mineralisation process is completed and N compounds are oxidised into nitrate, this last giving a high absorbance at shorter wavelengths as already shown before. The resulting spectra thus show a great difference.

A last example (Fig. 21) concerns a more efficient wastewater treatment plant type integrating a denitrification step in the biological process. This is possible by using, for example, an anoxic zone before the aeration step and by recirculating the mixed liquor (wastewater and biological sludge) from the outlet of the aeration basin. In this case, the nitrate formed can be assimilated as a source of oxygen by specialised microorganisms and can be reduced into gaseous form.

Figure 21 shows the effect of a denitrification plant reducing the discharged nitrate concentration. The removal yields of these last two examples are generally better than for the first treatment plant (Table 3).

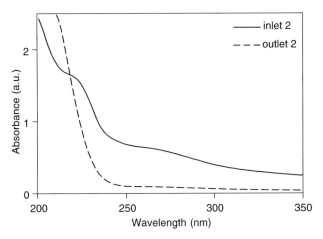

FIGURE 20. *Spectra of inlet and outlet of biological WWTP (low loading rate) outlet (diluted 2).*

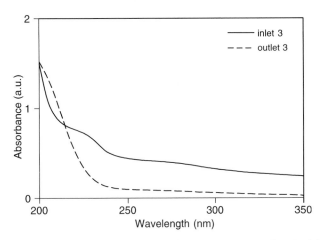

FIGURE 21. *Spectra of inlet and outlet of biological WWTP (denitrification) inlet (dilution 2).*

A first general comment is that the characteristics of treated wastewater obviously depend on the process installed in the treatment plant, but no one can totally remove the organic pollution. The efficiency, calculated in Table 3 for the whole treatment plant (integrating the efficiency of primary settling for TSS and related parameters), generally increases with the mean residence time of wastewater. The best results are obtained with the denitrification process, with more than 95% of yield on the different parameters. For the high-organic-load process, the mean efficiency is about 66% except for the TOC (34%). This is due to the fact that the biodegradation of organic compounds is partial as only a third of the organic carbon has been mineralised.

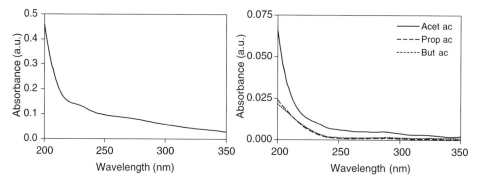

FIGURE 22. *Spectrum of outlet 3 without nitrate (see Fig. 21) and spectra of some carboxylic acids (acetic, propionic and butyric acids).*

Considering the differential spectrum of effluents corrected by the contribution of nitrate, the resulting shape can be explained by the presence of a few particles, residual organic compounds and small carboxylic acids as organic matter is not totally mineralised (Fig. 22). These compounds contribute to the remaining organic carbon.

3.4. Complementary technique: membrane filtration

Some complementary techniques can be used, such as membrane processes, for example. Urban wastewater treatment by membrane (e.g. microfiltration) can be envisaged up to a virtual disinfection, but the industrial development of these processes is still limited by the rather low value of the permeate fluxes and by membrane fouling that is not always reversible.

In order to have a better understanding of membrane fouling, UV spectrophotometry can be useful. Figure 23 presents the UV spectra evolution of a secondary effluent, from a biological treatment plant, before and after microfiltration. The decrease of the flux is due to membrane fouling. The stabilised flux is obtained after about 60 min of filtration.

The spectra of raw sewage and its filtrate show that some material is retained by the membrane, leading to membrane fouling. It seems, however, that the quantity of matter retained decreases as filtration time increases. Indeed, after 1 h of filtration, the quantity of matter retained by the membrane is lower than the one retained after 15 min of filtration.

This can be explained by the fact that particles retained by filtration are not only colloidal particles but also soluble matter that adsorb on the membrane. After 1 h of filtration, the membrane is so clogged that soluble matter cannot adsorb any more and, therefore, are no longer retained [15].

4. APPLICATIONS

Applications concerning the study of two wastewater treatment plants are presented. As the previous examples come from treatment plants with classical physico-chemical or biological processes (activated sludge), the two following examples have been chosen

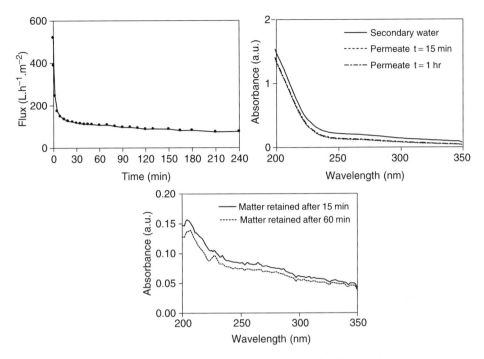

FIGURE 23. *Microfiltration of a biological treatment plant outlet. Permeate flux versus time, spectra of the effluent before and after filtration, and spectra of the matter retained by the membrane depending on the filtration time.*

for their particular design. The first is a coupled biological process integrating both a trickling filter and a biofilter. This treatment plant type is rather rare, but its efficiency is quite good. The second application is an extensive process, comprising several lagoons (aerated or not). This design is very interesting for economical reasons, particularly in sunny (and dry) areas. These examples are very different; on one hand, because of the process types, one intensive (short residence time) and the other extensive (several days of treatment); and on the other hand, by the management mode of degrading biomass.

4.1. Fixed biomass treatment plant

This wastewater treatment plant of a medium-size urban area (about 200,000 inhabitants) is located at Nimes, in the South of France. It includes two biological processes with fixed biomass, with a clarifier in between. The first step is a trickling filter using plastic material, and the second one is a biological filtration on immersed bed (with pouzzolane).

Several instantaneous samples have been taken along the treatment, at the inlet, after pretreatment (screening), after the first biological step, after the clarifier and after the biofilter corresponding to the outlet of the plant.

The acquisition of UV spectra has been quickly made, and the results are presented in Fig. 24. The raw wastewater has a regular shape, and the other spectra give some

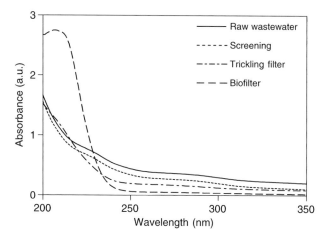

FIGURE 24. *UV spectra evolution during biological treatment (dilution 5).*

information on the characteristics of wastewater along the treatment, such as the degradation of organic matter and the nitrate formation with the biofiltration step. The spectra interpretation must take into account that the sampling mode (grab samples) is not relevant with the aim of the experiment (integrated sampling should have been preferable). From these data, the study of differential spectra (Fig. 25) gives another type of information.

More precisely, the spectrum evolution shows that the effect of screening on wastewater is equivalent to a primary settling. Coarse particles, the size of which is above 100 μm, are removed, and the resulting UV spectrum presents a diffuse shape due to the presence of supracolloids and colloids not removed by physical treatment.

An important part of organic matter is removed in the trickling filter, and a slight nitrate formation can be deduced from a decrease of absorbance after 240 nm and a small

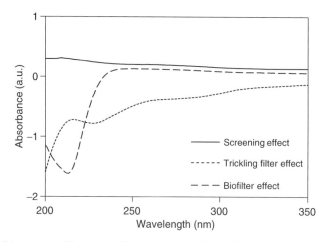

FIGURE 25. *UV spectra differences (from spectra of Fig. 24).*

TABLE 4. *Parameters evolution during treatment*

	Raw	Screening	Trickling bed	Biofilter
pH	7,8	7,7	7,3	7,1
TSS (mg/L)	220	75	60,5	15
BOD$_5$ (mg O$_2$/L)*	355	256	160	<20
COD (mg O$_2$/L)	577	416	192	48
DOC (mg/L)	57	57	31	19
IC (mg/L)	70	70	74	31
Surfactants (mg/L)*	11	10	<5	<5
NO$_3^-$ (mg/L)*	0	0	10	68

*Estimated by UV.

increase of absorbance between 200 and 210 nm with a Gaussian shape characteristic of nitrate. The shoulder at 225 nm has disappeared, showing that a great part of anionic surfactants have been removed.

The nitrification and organic matter removal is completed in the biofilter. Residual absorbance after 240 nm is close to zero, showing a very good efficiency of the process on organic matter degradation, and the Gaussian shape below 240 nm indicates a good nitrification. The formation of nitrate is obvious and shown either from raw spectra shape or from the differential spectrum. The difference between raw wastewater UV spectrum and treated wastewater is negative, demonstrating nitrate creation.

The measurement of aggregate parameters confirms these observations (Table 4). During the first biological step, 67% of COD and TSS are removed with 55% of BOD$_5$. However, most of the surfactants are degraded. The final biological step improves the global efficiency by more than 90% for TSS, BOD$_5$ and COD, but the DOC removal is equal to 67%, showing that residual organic by-products remain (mainly carboxylic acids). The decrease of hydrogenocarbonates is due to nitrification bacteria (autotrophs).

4.2. Extensive process

The treatment plant of Meze (more than 20,000 inhabitants) is located in the South of France, near Montpellier. It is made up of 11 lagoons or ponds of different volumes and functions. After two deep lagoons, the first four ponds are aerated, followed by three nonaerated (natural) ponds and by two final lagoons for polishing. The residence time of wastewater is about 2 months. The surface area of the lagoons varies from 15,000 to 39,000 m^2, and the average depth is 1 m. An experiment has been carried out in summer 1999. Eleven sampling points corresponding to the main steps of the treatment plant have been chosen (Fig. 26).

Figure 27 shows the evolution of UV-visible spectra of wastewater with treatment. The choice of considering the visible region for this application is related to the phenomena involved in this type of extensive treatment. In addition to the biodegradation process occurring either in suspended biomass or in sediments, the design of basins of weak depth allows working with direct or indirect sun radiation effects. In contrast to the other processes, the evolution of spectra shows an increase of absorbance for the whole spectrum, but is limited to the first steps. This is due to the existence of suspended

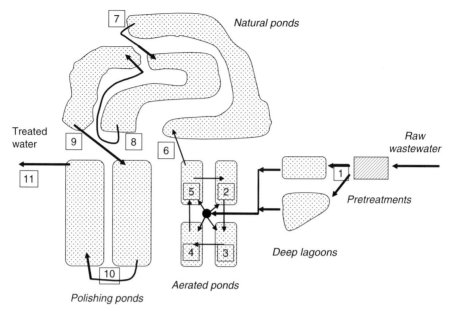

FIGURE 26. *Extensive treatment plant of Meze, and sampling points.*

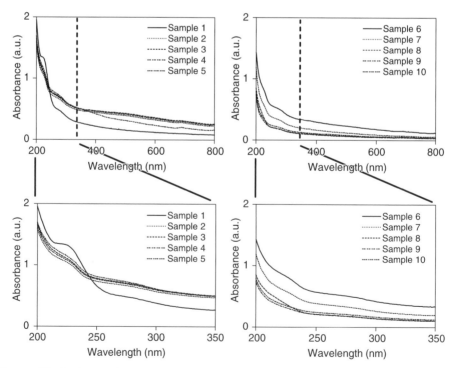

FIGURE 27. *Evolution of UV-visible spectra of wastewater during extensive treatment steps (dilution 5). Up, spectra between 200 and 800 nm; down, spectra between 200 and 350 nm.*

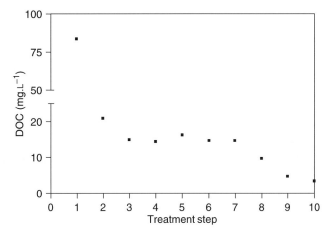

FIGURE 28. *DOC evolution with the extensive treatment efficiency.*

biomass, particularly in the aerated basins. Thus, the effect of biodegradation is clearly shown in the first step with the removal of most of surfactants (both partially degraded and adsorbed on the biomass floc).

Then, the spectra shape decreases up to the final lagoon, but shows a residual diffuse absorbance for all steps. The study of UV absorbance evolution, as well as the one of DOC concentration (Fig. 28), confirms these observations and shows that anthropogenic organic matter is still present up to the entry of final lagoons.

The evolution of TSS spectra in the second part of the treatment shows (Fig. 29) that the TSS nature changes from a classical composition (already studied previously) with the presence of organic compounds adsorbed on particles to another one characterised by a more featureless UV spectrum shape with an increase for the shortest wavelengths.

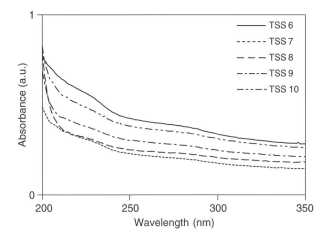

FIGURE 29. *Spectra of TSS retained in the last lagoons.*

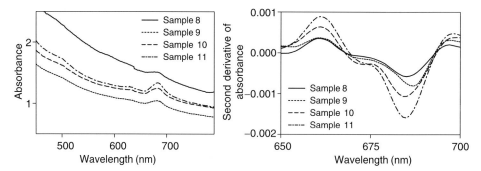

F<small>IGURE</small> 30. *Visible spectra (and second derivative) of nondiluted samples for the final lagoons (pathlength 50 mm).*

Considering that the last part of the treatment is characterised by the presence of microphytes (plankton), confirmed by the visual observation of the sample, and obviously by the visible spectra of samples, it is interesting to focus the study on the evolution of treated wastewater quality at the end of the treatment. The study of the visible spectra of samples of the three last lagoons shows a specific absorbance related to the presence of chlorophyll a, with a peak at around 680 nm. This peak, shifted to 664 nm in an acetone extract, is used for the determination of chlorophyll [16]. Notice that the use of the second derivative is interesting (Fig. 30). The presence of microphytes can be explained both by the weak depth of basins allowing the penetration of solar radiation and by the concentration of nutrients in treated wastewater (nitrate and phosphates) rapidly assimilated. In these conditions, photosynthesis is possible, leading to a correlative consumption of the residual organic matter used by the biocenose. The resulting global DOC removal is more than 95%, and this efficiency must be compared to the one of intensive processes, such as the previous example (67%).

5. CLASSIFICATION OF WASTEWATER

All the previous examples show that there is determinism between the shape of UV spectra and the evolution of wastewater quality with treatment. It is thus possible to try to propose a wastewater classification based on the UV characteristics. Two approaches can be proposed. The first is to find the most frequent spectrum types encountered and to give useful information to the user, and the second approach is to give a more automatic method based on the semi-deterministic method of spectra exploitation.

5.1. Typology of urban wastewater from UV spectra shape

Starting from the experience and the previous examples, four groups of urban wastewater can be proposed (Fig. 31). One group concerns the raw wastewater often characterised by the presence of shoulders at around 225 and 260–270 nm. The general spectrum is important, regularly decreasing, and a dilution of the sample is often needed in order to prevent absorbance saturation, particularly for short wavelengths. The shoulders can be explained by the presence of surfactants (of benzenic type) and of other anthropic

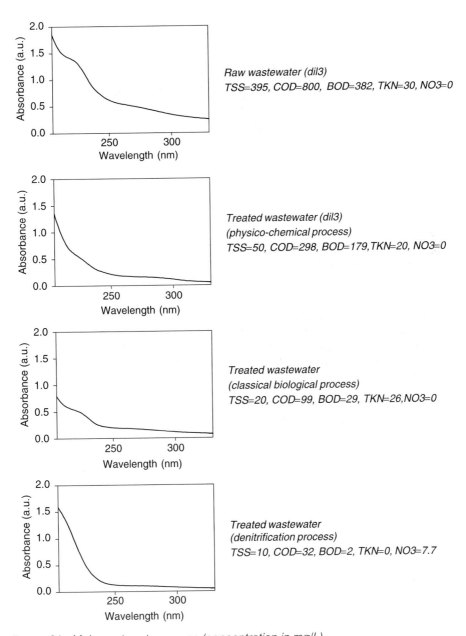

Raw wastewater (dil3)
TSS=395, COD=800, BOD=382, TKN=30, NO3=0

Treated wastewater (dil3)
(physico-chemical process)
TSS=50, COD=298, BOD=179,TKN=20, NO3=0

Treated wastewater
(classical biological process)
TSS=20, COD=99, BOD=29, TKN=26,NO3=0

Treated wastewater
(denitrification process)
TSS=10, COD=32, BOD=2, TKN=0, NO3=7.7

FIGURE 31. *Main wastewater groups (concentration in mg/L).*

TABLE 5. *Coefficient values from the deconvolution method for various types of wastewater and surface water*

	Type	Coefficient values					
		a1	a2	a3	a4	a5	(a1+a2+a3+a4+a5)
Raw wastewater		4.60	6.00	4.20	0	0.03	15.10
Treated wastewater	Chemical treatment	0.05	8.44	3.36	0	0.07	11.92
	Biological treatment	1.64	0.50	0.92	0.19	0.05	3.11
	Nitrification plant	0.50	0.38	0.54	2.88	0.02	1.44
	Denitrification plant	0.18	0.25	0.58	0.47	0.01	1.02
Surface water (receiving medium)		0.28	0.80	0.24	1.18	0.04	1.36
Surface water (without anthropogenic pollution)		0.02	0.03	0.02	0.06	0.00	0.07

organic compounds, often containing aromatic structures. The residual absorbance in the 320–350 nm regions is due to suspended solids.

The three other groups are concerned with treated wastewaters. The second type is wastewater treated by physico-chemical process. The resulting spectrum shape is decreasing, with no specific feature. The spectrum remains important as the suspended solids and supracolloidal fraction are only removed. The shoulder specific of the presence of surfactants is lowered as they are partially adsorbed on solids. The third type is related to classical biological treatment. Depending on the mean residence time and on organic loading, the spectrum keeps its initial shape, but its decrease is more or less important. The biodegradation process seems to affect all compounds of the readily degradable fraction. The last group concerns wastewater that contains nitrate. In this case, the organic matrix is becoming very simple with the presence of residual carboxylic acids. The nitrate concentration can vary and reach more than 50 mg/L, but can be reduced to a few mg/L if a denitrification process exists.

5.2. Automatic classification of water and wastewater

This automatic classification is based on the value of the coefficient contribution of the reference spectra used in the semi-deterministic method (see Chapter 2). The reference spectra are related to suspended solids, colloids, dissolved organic compounds, surfactants and nitrate. For the determination of the coefficient values, an important experiment has been carried out [17]. The deconvolution procedure was applied, on one hand, to several raw or treated samples taken at the inlet and outlet of several hundreds of wastewater treatment plants of different types, and, on the other hand, to a lot of surface water samples polluted with treated wastewater discharges. Table 5 shows the coefficients of the reference spectra for some of these samples (the coefficient values correspond to samples without dilution). The interpretation of the coefficients can lead to a rapid characterisation of the effluent type (inlet or outlet) and of the treatment plant type (chemical, biological, etc.).

Raw wastewater is characterised by high values of the first three coefficients (corresponding to suspended, colloidal and dissolved matter). The sum of coefficients is also

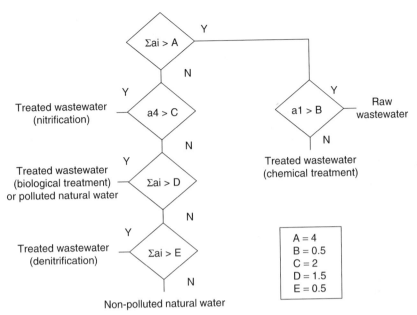

FIGURE 32. *Method for the classification of water and wastewater based on the values of reference UV spectra contribution.*

very high for raw wastewater, as well as for the outlet of a chemical treatment plant. The latter is characterised by a high concentration of colloidal and dissolved matter (due to the process) but by a very low contribution of suspended matter. For a treated sample from a biological treatment plant, the sum of the coefficients (except the fourth, corresponding to nitrate) is much lower than for the former samples. Indeed, the coefficients generally decrease as the quality improves, except for the contribution of nitrate, the maximum of which is obviously related to the nitrification process. For surface water, all coefficients are close to zero.

From the coefficient values, a generalisation can be proposed for the classification of wastewater (Fig. 32). This classification is based both on the value of the sum of the coefficients, excluding the fourth one corresponding to nitrate, and on the magnitude of the first (TSS) and fourth (nitrate) coefficients. This method can be applied in order to show the presence of anthropogenic organic compounds in natural water or to check the efficiency of a chemical or biological treatment.

References

1. N. Ogura, T. Hanya, *Nature*, 212 (1966) 758.
2. M. Mrkva, *Water Res.*, 9 (1975) 587.
3. R. Briggs, J.W. Schoefield, P.A. Gordon, *Wat. Pollut. Control Fed.*, 75 (1976) 47.
4. S. Vaillant, *Organic Matter of Urban Wastewater: Characterisation and Evolution*. PhD thesis, Université de Pau et des Pays de l'Adour, France (2000).
5. S.R. Qasim, *Wastewater Treatment Plants Planning, Design and Operation*, Technomic Publishing Co, Inc, Lancaster, Basel, 1994.

6. M-F. Pouët, C. Muret, E. Touraud, S. Vaillant, O. Thomas, O. *Proceedings of Interkama-Isa Conference on CD ROM*, Düsseldorf, Germany,1999.
7. F. Valiron, J.P. Tabuchi, *Tec & Doc*, Lavoisier, Paris, 1992.
8. S. Michelbach, *Wat. Sci. Tech.*, 7 (1995) 31.
9. M. Desbordes, *Journée Eau et Circulation des Polluants en Pays Méditerranéen*, Pôle Universitaire Européen, Montpellier, France, 1995.
10. S. Vaillant, M.-F. Pouet, O. Thomas, *Talanta*, 50 (1999) 729.
11. S. Gallot, O. Thomas, *Fres. J. Anal. Chem.*, 346 (1993) 976.
12. S. Vaillant, M.-F. Pouet, O. Thomas, *Proc. Novatech* 3rd *Int. Conf. Innovative Technologies in Urban Storm Drainage*, Lyon, France, 1 (1998) 39.
13. F. Theraulaz, L. Djellal, O. Thomas, *Tenside Surf. Det.*, 33 (1996) 447.
14. L. Djellal, F. Theraulaz, O. Thomas., *Tenside Surf. Det.*, 34, (1997) 316.
15. M-F. Pouët, A. Grasmick, in *Proc. Euromembrane '95 Bath*, ed by W. Richard Bowen, Robert W Field and John A Howell (1995) 482.
16. A.K. Bhadra, *Conference of the American Chemical Society*, San Francisco, 26–30 March, 2000.
17. O. Thomas, F. Theraulaz, C. Agnel, S. Suryani, *Environ. Technol.*, 17 (1996) 251.

CHAPTER 9

Industrial Wastewater

O. Thomas[a], H. Decherf[b], E. Touraud[c], E. Baurès[a], M.-F. Pouet[a]

[a]Observatoire de l'Environnement et du Développement Durable, Université de Sherbrooke, Sherbrooke, Québec, J1K 2R1, Canada; [b]Veolia Eau, Agence Industrie Berre Provence, 13857 Aix en Provence Cedex 03, France; [c]Laboratoire Génie de l'Environnement Industriel, Ecole des Mines d'Alès, 6 Avenue de Clavières, 30319 Alès Cedex, France

1. INTRODUCTION

An attempt towards the UV spectrophotometric study of industrial wastewater can be considered unrealistic because, on the one hand, of the potential presence of nonabsorbing compounds and, on the other hand, of the huge variety of effluents released by industrial units very different in their activities and products. However, as industrial effluents are of a complex nature, nonabsorbing compounds are often accompanied by other absorbing substances, probably explaining that more than 95% of industrial wastewater samples present a rather well-structured UV spectrum. Moreover, as will be seen in this section, a typology of industrial wastewater UV spectra can be proposed, taking into account the presence of pollutants and their degradation along sewer and treatment systems.

In a first part, the UV spectra of raw industrial effluents are examined with respect to the industrial activity. Then, some practical qualitative and quantitative applications, mainly concerning petrochemistry, are proposed. These applications will show the interest of procedures developed for process assistance or treatment control. Concerning this last point, all real examples presented in this section are taken from French and U.S. industries, collecting their effluents with sewers including physical treatment tanks (settlers, flotators, etc.), before generally a unique treatment plant. In some countries, the legislator may oblige the industries to treat wastewater from all units before mixing, leading to an increase in the number of possible applications of UV spectrophotometry in wastewater quality control.

2. WASTEWATER CHARACTERISTICS

2.1. Generalities

Raw industrial effluents can be classified according to the dominant nature of pollution, organic or mineral (Table 1), and may be characterised by a high concentration of organic (and mineral) compounds, due either to a few major pollutants (chemical industry effluents, for example), or to a huge number of molecules, the concentration of which is very low (pulp and paper or food industry).

TABLE 1. *Types of industrial wastewater*

Type	Characteristics	Industry (example)	Treatment
Organic	High organic pollution, easily biodegradable	Food industry	Biological simple
Organic	High organic pollution, not easily biodegradable	Refinery Petrochemistry	Biological adapted
Organic	High organic pollution, nonbiodegradable	Organic synthesis	Physico-chemical Biological adapted
Mineral	Low organic pollution Toxics High suspended solids	Steel industry Electroplating industry Extractive industry	Physico-chemical
Miscellaneous	Organic pollution (with major pollutants), high salinity	Chemistry	Physico-chemical Biological adapted
Miscellaneous	Organic pollution (mixture), salinity	Pulp and paper industry	Physico-chemical Biological adapted

Raw industrial wastewater is generally produced continuously during the industrial process, but can include some other liquid wastes such as washing residues and sometimes process water, in case of an incident. For these reasons, all industrial effluents may vary in quality and, obviously, in the corresponding pollution load [1,2].

2.2. Influence of industry nature

One way of studying the influence of industry nature on the shape of UV spectra is to consider the organic fraction. Firstly, food industry wastewater is generally characterised by a high content of organic biodegradable pollution coming from the process of natural matrices, but also of suspended solids, these being easily retained by a physical treatment. The corresponding UV spectra present a featureless variation with no evident peaks but with a shoulder at around 275 nm (Fig. 1). The general shape is conserved from one food industry to another, but this observation must be validated for other types of food industries (spectra of food dye or drinks units will probably be different).

Another kind of industrial activity processing natural products is the pulp and paper industry, the wastewater quality of which is characterised by a rather high concentration of suspended solids (such as, for example, residual cellulose fibres) and of organic non-easily-biodegradable load, depending on the process chosen for pulp production (chemical or semi-chemical). The corresponding UV spectra are generally featureless, similar to food industries with a slight shoulder around 280 nm, probably because of phenolic compounds, and a noticeable residual absorbance after 300 nm due to suspended solids (Fig. 1).

Contrary to the two first classes of industrial wastewater, the chemical industry, and particularly the organic synthesis units, gives the most structured UV spectra for wastewater (Fig. 1) as far as the molecules synthesised are complex, as, for example, for pharmaceutical units. Taking into account the huge variety of molecules synthesised nowadays, it is impossible to show some characteristic examples of UV spectra of wastewater. Therefore, each industry has to be considered as unique with its own potential applications.

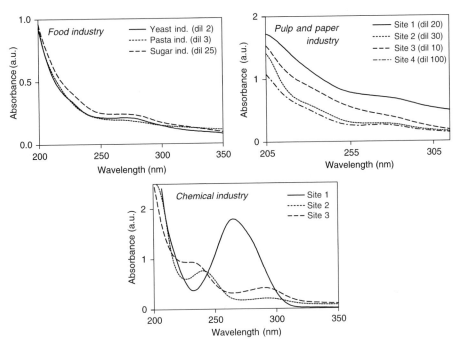

FIGURE 1. *Example of UV spectra of food, pulp and paper and chemical industries.*

Actually some chemical activities as refineries are not so variable on the point of view of wastewater quality because of the large quantities of primary matter and products (Fig. 2). Near units, the UV spectrum shape often shows peaks due to the presence of additives used in the process, acting as revelatory of the organic pollution load [3]. Without these compounds and considering that saturated organic compounds do not absorb, it would be difficult to use the UV signal for wastewater quality control. However, a nonparametric measurement (NPM) based on comparison of the UV absorption spectra can be used for the characterisation of wastewater quality variability [4]. Obviously, as far as wastewater

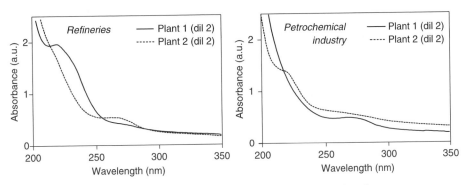

FIGURE 2. *Example of spectra of refineries and petrochemical industries.*

TABLE 2. Coarse classification of industrial wastewater based on UV spectrophotometry

Spectrum characteristics	Industry	Observations
Structured (peaks)	Refinery Petrochemistry Chemistry (organic synthesis)	Major pollutant(s)
Featureless	Food industry Pulp and paper	Suspended solids

from industrial units is mixed along sewers, the general shape becomes featureless at the inlet of the treatment plant as it will be shown afterwards.

For the petrochemical industry, the characteristics and evolution of UV spectrum shape observed for refineries are conserved, however, with best-defined peaks. These last are explained by the presence of additives, intermediates and other chemicals (Fig. 2).

At the end of this part, according to some characteristics of UV spectra (peaks, shoulders or residual absorbance at high wavelengths), a tentative proposal of the typology of raw industrial wastewater can be made (Table 2). This coarse classification is mainly related to the presence of major organic pollutants in the sample. Other parameters for organic pollution characterisation (total organic carbon, TOC; chemical oxygen demand, COD; and biological oxygen demand, BOD), for physico-chemical measurement (suspended solids conductivity, pH, etc.) or for specific-compound determination can obviously complete the classification.

2.3. Variability of industrial wastewater

This point is very important mainly in view of the design of a measurement strategy for industrial wastewater quality control. Among the analytical techniques available for the study of the complexity and qualitative variability of the medium, UV spectrophotometry is well adapted for the quality variation control of industrial wastewater (see Chapter 2).

An example is given in Fig. 3 showing the spectra of different effluents, sampled from the inner sewer of a refinery. The first observation is that all spectra are structured with at least one peak in the 250–300 nm region due to the presence of cyclic compounds such as phenols. The second one is that some fluxes present an absorption band around 230 nm, related to the presence of sulphur compounds.

A more precise study has been made from samples of another refinery site. Figure 4 displays the two sets of spectra corresponding to the inlet of the wastewater treatment plant (samples 1) and to the main flux upstream at the outlet of the desalting unit (sample 2). The raw spectra do not present a direct isosbestic point (IP), but, after normalisation (see Chapter 2), the two sets are characterised by the presence of a hidden isosbestic point (HIP) [6], more or less evident, as shown on the zoom in Fig. 4. Considering the experimental error related to the absorbance measurement, the isosbestic point may rather be considered as an isosbestic surface.

An interesting application of the use of HIP, the significance of which has been discussed in Chapter 2 (quality conservation), is the estimation of the variability V of an effluent according to the following relation [7]:

$$V = \left(1 - \frac{Npi}{Nt}\right) \times 100$$

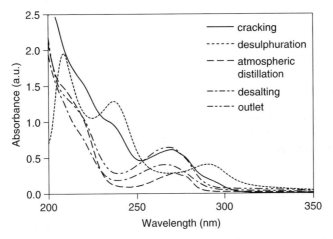

FIGURE 3. *UV spectra of process waters in a refinery (pH = 11, pathlength: 10 mm, dilution: 10) [5].*

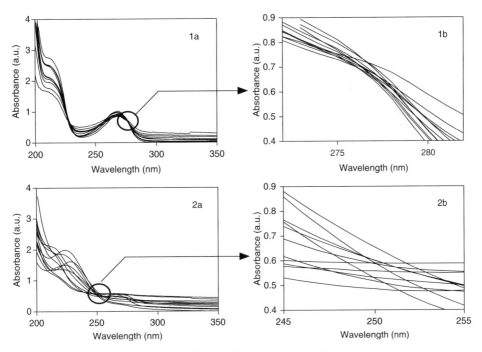

FIGURE 4. *Use of HIP for the variability study (example of refinery wastewater (a) inlet of treatment plant, (b) outlet of desalting unit) [4].*

where Nt is the total number of spectra, and Npi, the number of spectra crossing at the HIP (or direct IP if any).

Taking into account this approach, the two effluents present some differences (Fig. 4). Sample 1 (inlet of wastewater treatment plant) is characterised by an HIP at 276 nm, with a variability of 13%; meanwhile, sample 2 has an HIP at 250 nm with a variability of 67%. These results can be explained by the "buffer" effect of the fluxes mixing at the inlet of the treatment plant. This can be generalised for all industrial plants, the effluent variability decreasing from the process units to the treatment plant.

This qualitative parameter (V) is particularly interesting for the survey of a network (incident detection, connection error, etc.) and for the optimisation of the wastewater treatment plant design, in tandem with quantitative parameters.

2.4. Quantitative estimation

One of the most important applications of UV spectrophotometry, particularly for industrial wastewater quality control, is the rapid estimation of the concentration of some substances or some parameter values, among which are aggregate organic parameters and N or P compounds (Table 3).

In the case of industrial wastewater, several other compounds can also be involved, such as sulphide, hexavalent chromium or organic molecules. This last group of compounds is considered hereafter with regard to their economic and/or environmental importance. For example, Fig. 5 presents several specific compounds encountered in refinery and petrochemical wastewater. Thus, phenol, EPA (ethylpropylacrolein), TBC (tertiobutylcatechol), NMP (N-methylpyrolidone) and nitrite can be detected in effluents or process water [21]. Moreover, the estimation of complementary aggregate parameters, such as total oxygen demand (TOD), is possible from the estimation of one of the previous organic compounds [3] (Fig. 6).

The semi-deterministic method (already described, see Chapter 2) has been applied for the survey of treated wastewater of several different industries (chemical, petrochemical and electroplating). In some cases, when the UV response of the matrix is very

TABLE 3. Quantitative applications of UV spectrophotometry for industrial wastewater

Parameters	Applications	Principle	Ref.
COD, TOC, BOD5, TSS	Urban, petrochemical, paper, distillery effluents	Deconvolution	[2,8,9]
TSS, colloids	Coagulation-flocculation process assistance, pulp and paper and agrofood effluents	Deconvolution	[10–13]
Nitrates	Urban, petrochemical, paper, distillery effluents	Deconvolution	[14]
Organic and ammonium nitrogen and phosphorus	Urban, petrochemical, paper, distillery effluents	UV/UV system, deconvolution	[15]
Sulphide and mercaptans	Petrochemical effluents	Deconvolution	[16,17,9]
Hexavalent chromium	Electroplating effluents	Deconvolution	[18]
Surfactants	Urban effluents	Deconvolution	[19]
Phenols and photodegraded compounds	Chemical effluents	UV/UV system, deconvolution	[20]

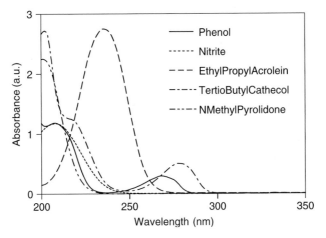

FIGURE 5. *UV spectra of specific compounds in refineries.*

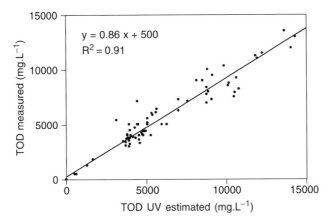

FIGURE 6. *Comparison between measured and estimated values of TOD in petrochemical wastewater (the estimation is made from the ethylpropylacrolein determination; see Fig. 5) [3].*

important, the working wavelength range must be modified in order to allow parameter calculation. This is the case for the survey of electroplating wastewater, where nitrate concentration is very high and chromium (VI) concentration rather low (Fig. 7).

Table 4 shows the correlation between the results obtained with standard methods and with UV spectrophotometry. The number of samples used for comparison is greater than 50 for all parameters. Except for TSS estimation, the determination coefficient values are close or greater than 0.9. These quantitative results, quite good for a rapid and direct method, are explained by the fact that the calculation is validated only if the quadratic error is lower than an accepted one (for example, 1%). If not, for about 10% of the samples, the result is not displayed but the information is interesting and will be exploited earlier.

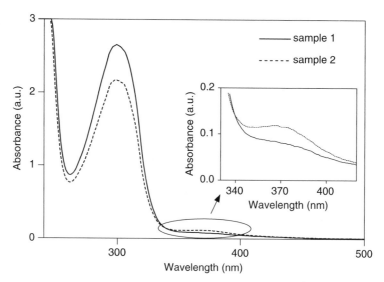

FIGURE 7. *UV spectra of electroplating effluents: (a) with 6.6 g.L⁻¹ of nitrate and (b) with 5.4 g/L of nitrate and 450 µg/L of chromium (VI).*

TABLE 4. *Correlation between UV spectrophotometry results and industrial water quality parameters*

Parameters	Industry	Range*	R^2
TOC	Petrochemical	5–60 mg/L	0.91
COD	Petrochemical	5–150 mg/L	0.89
TSS	Petrochemical	10–100 mg/L	0.77
NO_3^-	Chemical	1–50 mg/L	0.99
Cr IV	Electroplating	10–300 µg/L	0.96
Phenols	Petrochemical	5–100 mg/L	0.90
Sulphide (HS⁻)	Petrochemical	0.5–15 mg/L	0.95

*For a pathlength of 10 mm, without dilution.

3. TREATMENT PROCESSES

As for the treatment of urban wastewater, two main treatment types are used, based on physico-chemical or biological principles.

3.1. Physico-chemical processes

Physico-chemical treatments include either separation techniques or processes based on a chemical reaction. As decantation and filtration have already been studied before for urban wastewater or for natural water, only complementary processes, sometimes largely used for industrial wastewater, are presented in this section.

A more simple treatment is the pH correction, very often used for regulation compliance (the pH value of the treated effluent must be between 5.5 and 8.5). A pH modification can also be used for metallic compound precipitation usually as hydroxide forms, in alkaline conditions, or for humic substances removal, in acidic conditions. The effect of this last treatment can be shown for landfill leachates treatment in Chapter 10.

Coagulation–flocculation is very often proposed for the removal of colloidal fraction and fine particles. The UV study of the coagulant concentration effect (FeCl$_3$, for example) has already been presented in Chapter 8.

Another treatment, currently used for industrial wastewater, is the adsorption on active carbon for organic pollutants removal. UV spectrophotometry can be proposed for the study of the effect of granular active carbon (GAC) on the adsorption of organic compound of a chemical effluent (Fig. 8). This example shows that the molecule, characterised by an absorption peak at 238 nm (not identified), is well adsorbed; meanwhile, the one absorbing at 260–270 nm is not retained. The corresponding removal rates of TOC are, respectively, 27, 36 and 51% for the three GAC concentrations (5, 10 and 20 g.L^{-1}). UV spectrophotometry can thus be used for process control and for the quality monitoring of the treated effluent.

Before considering the interest of UV spectrophotometry for the control of biological processes, a last physical treatment type must be presented. Advanced oxidation processes (AOP) are more and more used, because of their destruction power, preventing the pollution transfer (as it is the case for the other processes). AOP schemes include ozonisation, photo-oxidation and photo-catalysis processes, these last being based on the effect of UV radiation.

Figure 9 presents the effect of some AOPs on the treatment of coloured wastewater from textile industry. Two tests have been carried out, the first with UV radiation alone (photodegradation) and the second one with UV and hydrogen peroxide (0.1 mol/gCOD) [22]. The results are compared to those from the existing biological treatment plant. The gain for the AOP solutions is clearly demonstrated, particularly for the UV peroxide process. The discoloration is very rapid, and the corresponding TOC removal rate is about 90% for 15 min of treatment.

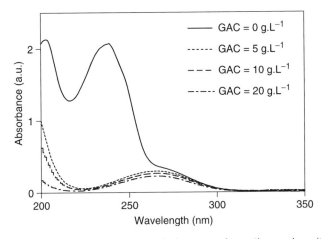

FIGURE 8. *Spectra of chemical wastewater during granular active carbon (GAC) adsorption tests (TOC values are, respectively, 1373, 1001, 877 and 670 mg.L^{-1}).*

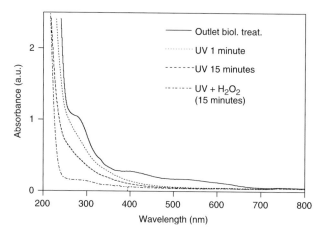

FIGURE 9. *Comparison of AOP and biological process for the treatment of coloured textile wastewater.*

3.2. Biological processes

Some examples of industrial biological treatment plant are presented. Despite the high pollution load of some industrial wastewater, such as for chemical and petrochemical industry, for example, and the presence of non-easily-biodegradable organic compounds (in some cases, BOD_5 cannot be measured), biological processes are often able to treat this type of pollution. In this case, the biomass is "naturally" acclimated (microorganisms of biological reactors), but nutrients (N and P compounds) have to be added if needed.

   The first example (Fig. 10) shows the evolution of UV spectra of raw and treated wastewater for one refinery and one petrochemical site, with two different biological treatment plants, one with fixed biomass, and the other with activated sludge. For both cases, the treatment efficiency is good, with a TOC removal of around 90% (either measured or UV-estimated) and the presence of nitrate in the treated effluent.

   Another example is related to the biological treatment of pulp and paper wastewater (Fig. 11). The shape of raw wastewater is less structured than for chemical

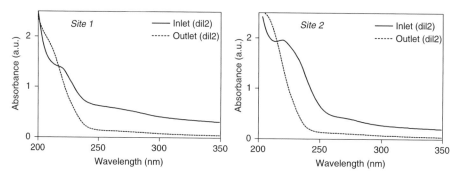

FIGURE 10. *Treatment plants efficiency for refinery (site 1) and petrochemical (site 2) wastewater.*

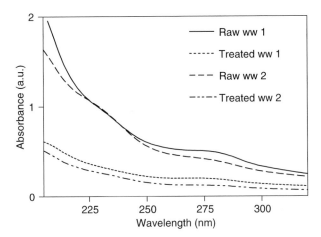

FIGURE 11. *Effect of biological treatment on UV spectra of pulp and paper wastewaters (ww 1 and 2).*

or petrochemical industry as previously seen, and the effect of biological treatment is characterised by the presence of a residual absorption on the spectrum of the treated wastewater, probably due to the remaining fine cellulose particles. Notice that, in contrast to the last example, no nitrate is "visible". The TOC removal rate is less (85%), and the raw effluent contains few N compounds (particularly organic N).

3.3. Hyphenated processes

Generally, industrial treatment plants integrate hyphenated processes with pretreatment steps more important than for urban wastewater treatment. In this case, UV spectra allow showing the effect of the different treatment processes.

A first example is shown on Fig. 12 for a treatment plant of a refinery (different from the one in Fig. 10), including a separator of hydrocarbons (API tank) and a sand filtration unit before a trickling filter. The raw wastewater composition is characterised by the presence of phenolic compounds with an absorption peak around 265 nm. The effect of the two pretreatment steps is evident on the particulate fraction (including the effect of emulsified hydrocarbons), but does not affect the dissolved matrix. This last is removed (at least the phenolic compounds) with the biological step, the effect of which is a removal of almost 90% of the TOC. At the end of the treatment, cooling water (pumped from the sea) is mixed with treated wastewater, explaining the nitrate dilution and the presence of chloride in the discharge.

Figure 13 presents the raw and treated spectra of a coloured textile wastewater already studied in Fig. 9. The treatment plant includes a physico-chemical step and a biological process (activated sludge). The UV-visible spectra show that the first step leads to the removal of compounds responsible of at least 50% of wastewater colour. The effect of the biological step completes the TOC removal up to 80%, without degrading aromatic amines, responsible for the 270 nm shoulder. This well-known problem is easily shown with the use of UV-visible spectrophotometry.

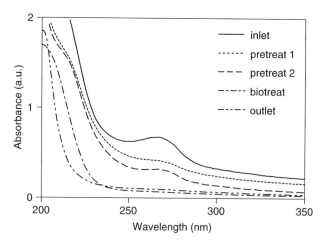

FIGURE 12. *Spectra of refinery wastewater during treatment (dilution twice).*

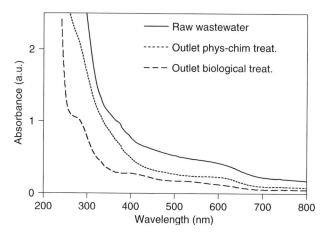

FIGURE 13. *Spectra of raw and treated textile wastewater.*

Since the relation between the UV spectrum shape and organic content (qualitative and quantitative) is evident in the previous industrial cases, a comparison concerning two ways of calculation of the efficiency of the organic pollution removal has been studied [3]. In Table 5, the results of TOC removal ratio and some UV yield estimations show that the TOC removal estimated from the UV semi-deterministic method is quite satisfactory.

4. WASTE MANAGEMENT

The first interest of UV spectrophotometry is to give some qualitative information such as the previous attempt of typologies for natural water and urban wastewater (see Chapters 7 and 8). From a more practical point of view, the study of UV spectra of industrial wastewater can be envisaged for several purposes such as sampling

TABLE 5. *Comparison of TOC removal yields in percentage (calculated or UV estimated)*
for different refinery treatment plants (site E has been studied several times)

Origin	TOC removal	Estimated TOC removal (semi-deterministic method)	Estimated TOC removal (A_{265})	Estimated (area under spectrum)
Site A	76.3	65.3	55.0	54.3
Site B	91.2	84.8	73.7	41.8
Site C	88.5	87.8	78.1	85.7
Site D	78.9	75.6	77.1	43.6
Site E (a)	87.2	78.1	37.2	65.0
Site E (b)	83.9	81.5	57.1	71.6
Site E (c)	89.3	87.5	73.5	80.7
Site E (d)	87.8	79.9	62.8	49.8

assistance, treatability study, wastewater quality control (incidents detection, waste characterisation, etc.), treatment control (efficiency, troubleshooting, etc.) and environmental
impact study [4].

4.1. Sampling assistance

The first application of wastewater management can be the checking of sample quality.
Figure 14 shows the spectra of different samples of the same effluent (taken from the
same location) of a petrochemical plant, one from a composite 24-h sampling procedure (with cold temperature conservation) and two from grab sampling, taken at 30-sec
intervals.

The spectra are very different, even for the two grab samples (differences due to suspended solids) and particularly between grab and composite samples. Two explanations

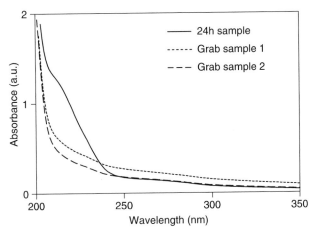

FIGURE 14. *Comparison of UV spectra of composite and grab samples of a*
petrochemical effluent.

can be advanced, a qualitative variation with time and the qualitative evolution of the composite sample during its storage. This example confirms that UV spectrophotometry can be used for the quality control of sample evolution [23,24] and for the study of the phenomenon of wastewater sample aging [25].

4.2. Treatability tests assistance

A treatability study is often needed for designing an industrial treatment plant. As biological processes are often preferred (because of their efficiency and simplicity to run), biodegradation tests are carried out. A sample of the industrial wastewater to be tested is introduced into a flask in the presence of microorganisms, generally coming from urban biological treatment plants. The evolution of BOD5 and COD is monitored during at least one day. Figure 15 presents the evolution, with time, of the UV spectra of the biodegradation test of a petrochemical effluent, showing the relative quick degradation of the main organic pollutant absorbing around 235 nm (within few hours).

Using the UV/UV photodegradation system used for the determination of some minerals (total nitrogen and phosphorous), as presented in Chapter 4, a photodegradation test can be proposed in order to reduce the time of the treatability study. The evolution of UV spectra with the photodegradation time can be compared to the one of the biodegradation test (Fig. 15). The evolution of spectra is very close, but much more rapid, as the same result seems to be obtained within few minutes. Studying the evolution of spectra more accurately, some slight differences can be seen, showing two ways of degradation leading to the degradation of the organic matrix.

A comparison of the two tests applied on different fluxes of petrochemical wastewater has been made [26], showing a quite good adjustment between the results (Fig. 16). This method, very interesting, might be carefully validated for other types of industrial wastewater, before being used extensively.

Another approach can be proposed in order to try to avoid the experimental step of degradation. Based on the relation between the light absorption of a substance and its chemical properties (see Chapter 1), the exploitation of the Shape Factor (presented in Chapter 2) is chosen for the purpose.

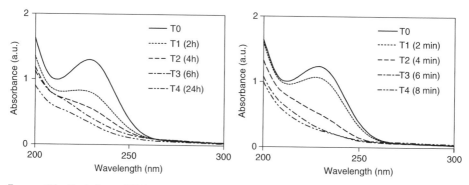

FIGURE 15. *Evolution of UV spectra during biodegradation and photodegradation tests of an industrial effluent.*

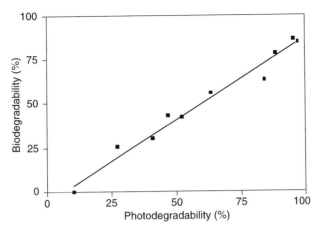

Figure 16. *Comparison between biodegradation and photodegradation tests for petrochemical effluents [26].*

The Shape Factor has been defined, for a peak, from the product of the ratio between the second derivative and the absorbance by the half width of the peak, as you may recall from Chapter 2. As this parameter does not take into account the position of the peak, the relative Shape Factor (RSF or SF*) is defined as follows:

$$SF^* = -\frac{D(\lambda)}{A(\lambda)} \times H \times \lambda$$

where $D(\lambda)$ is value of the second derivative measured at the wavelength λ, $A(\lambda)$ is the absorbance value measured at the same wavelength λ, and H, the width at the half height of the peak (of wavelength λ).

Before explaining the proposed procedure, several photodegradation tests have been carried out on different samples of industrial wastewater from chemical industries. Figure 17 shows the evolution of UV spectra for the different tests. The three groups of spectra (well, few and nonstructured; see Chapter 3) are represented. The evolution is different, depending on the initial spectrum shape. Two effluents of the first group (well-structured spectrum) present a direct isosbestic point, at least at the beginning of the test. This corresponds to the degradation of a major pollutant in the first minutes of the photodegradation experiment. The two others are characterised by a variation very low and even by no evolution. From these results, and particularly for the first minutes, it is possible to calculate a pseudo half reaction time ($t_{1/2}$) [27]. The results are presented in Table 6.

The presence of the wavelength of the peak in the relation tends to balance the SF* value by the energy associated with the peak (hC/λ). Indeed, it can be roughly demonstrated that a given molecule is more easily degraded than its spectrum presents one or several peaks in the higher wavelengths (in the visible region). The results presented in Table 6 show that an effluent characterised by a well-structured spectrum is more rapidly degraded than one with a featureless spectrum. This observation is at least right for the initial pollutants but has to be confirmed for the by-products of the degradation.

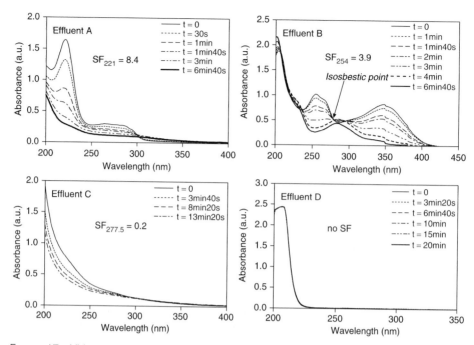

FIGURE 17. *UV spectra evolution during photodegradation tests (respective dilutions of 100, 5, 5, and 50 for samples A, B, C and D) [27].*

TABLE 6. *Characteristics of the industrial wastewater samples (Fig. 17) and SF values*

Effluent	Cond (mScm^{-1})	pH (upH)	COD (gL^{-1})	TOCi (gL^{-1})	$t_{1/2}$ (min)	SF (10^2)	SF*	λ (nm)
A	0.15	9.6	550	86.7	3.2	13	29	221
						8	24	287
B	0.1	7.5	20	5.3	0.4	5.5	15	254
						11	36	350
C	5.4	7.9	1	0.5	13.3	0.04	0.1	229
						0.2	0.5	277
D	185	13.1	6	1.6	>30	0	0	—

This approach shows that the prediction estimation of wastewater treatability seems to be possible from the direct study of UV spectra [26,28].

4.3. Spills detection

TOC monitoring is often used for industrial wastewater quality and sometimes completed by the TOD. Figure 18 presents the TOD variation with time (during almost 2 years) monitored at the inlet of the wastewater treatment plant of a petrochemical site [3]. Some peaks of concentration have been registered, and the chromatographic analysis of the

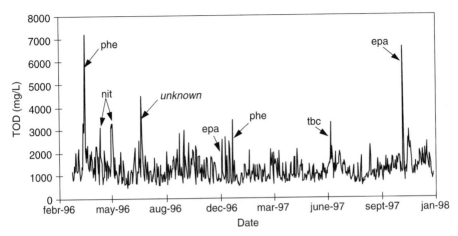

FIGURE 18. *Evolution of TOD and identification of major pollutants (phenol, nitrite, ethylpropylacrolein and tertiobutylcathecol).*

associated samples has permitted determining the major pollutant responsible of the high value of TOD (except once). However, considering that TOD measurement is sensible to interferences such as ammonia, TOC (by definition more adapted to the quantification of organic pollution; see Chapter 5) has been preferred. Thus, a modified monitoring phase has been carried out in the same site. For each TOC measurement (every 3 h), a UV spectrum has been acquired. Considering that the maximum admitted value for the protection of the biological reactor of the plant is 500 mg/L of TOC, three peaks of pollution have been detected (Fig. 19).

The study of the corresponding UV spectra (for TOC peaks) and a comparison with the one of regular wastewaters shows that these three incidents are related to three different accidental discharges in sewers. Moreover, the study of UV spectra shows that incidents 1 and 3 are characterised by the presence of a few concentrated pollutants. The UV spectra of effluents and those of main pollutants already mentioned before are presented in Fig. 20. Starting from this information, and after a more complete study, a UV monitor has been installed close to the main source of the potential accidental discharge. An automatic procedure (UVDIAG) for incident diagnosis and pollutant identification (if possible) has been included in the monitoring (see algorithm in Fig. 21).

4.4. Shock loading management

It is well known that a biological treatment plant is sensitive to pollution peaks, especially if the biomass is not adapted to the new substrate. This is the reason why storage tanks are often placed before the treatment plant for the regulation of organic loads. In this case, the choice of the location of monitoring tools (online UV analysers) is important in order to detect, as soon as possible, the abnormal variability of the effluent before its eventual storage. For this purpose, remote monitoring can be carried out (Fig. 22). Furthermore, outlet monitoring can be useful both for regulation compliance and post-treatment.

Outlet monitoring allows controlling the impact of the shock load on the treatment plant and, if necessary, to divert the outlet effluent into a storage tank for a further

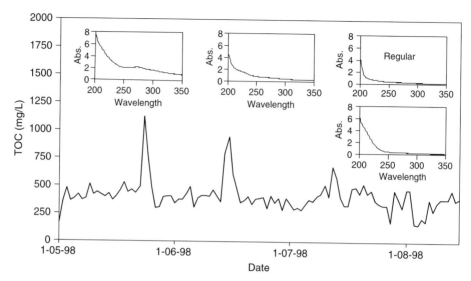

FIGURE 19. *TOC and UV monitoring of petrochemical wastewater (the absorbance values are for a 10-mm optical pathlength quartz cell).*

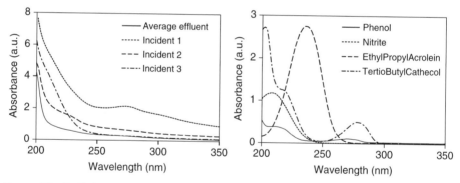

FIGURE 20. *Spill detection and related pollutants (the absorbance values are calculated from the dilution factor).*

treatment. The intention of remote monitoring is to prevent and point out a potential incident. Figure 23 illustrates the procedure to be applied in case of incident detection.

In case of shock loads, a procedure has been proposed and is schematised in Fig. 24. It takes into account the treatability diagnosis (photodegradation test), the management of an undetected incident on upstream effluent and the limits of UV spectrophotometry (nonabsorbing compounds).

Finally, it can be noticed in Fig. 24 that five cases can be observed according to the differences between inlet and outlet UV spectra, leading to different yields. The following applications will illustrate the different situations.

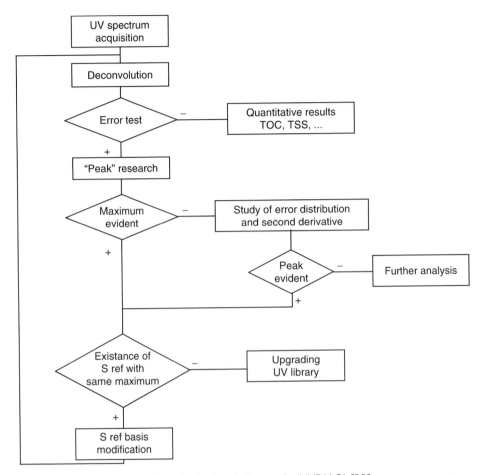

FIGURE 21. *General procedure for incident diagnosis (UVDIAG) [29].*

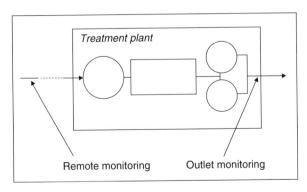

FIGURE 22. *Monitoring of biological treatment plant.*

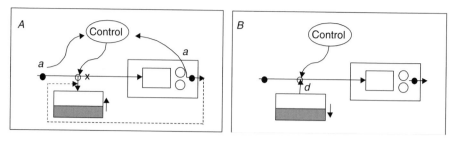

FIGURE 23. *General procedure for shock load management (A: detection and storage of the suspect effluent, a; B: treatment of effluent with a dilution factor, d).*

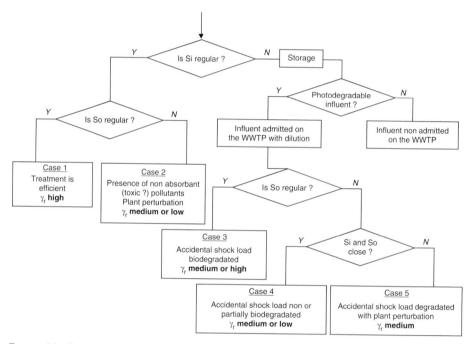

FIGURE 24. *Proposed procedure synopsis for shock load management (Si and So are the concentration (e.g. TOC) at the inlet and outlet of the plant, respectively, and γ_r the efficiency yield of the process).*

Figure 25 shows, for the different cases, several results (UV spectra comparison) from effluents sampled in chemical wastewater treatment plants. Table 7 gives the corresponding organic loads and removal yields.

Case 1 presents a significant absorbance decrease corresponding to high organic load abatement: UV spectra (Si and So) are regular.

Case 2 shows similar UV spectra, even if the yield is medium. It can be explained by the presence of nonabsorbing compounds that are partially degraded. The limits of the

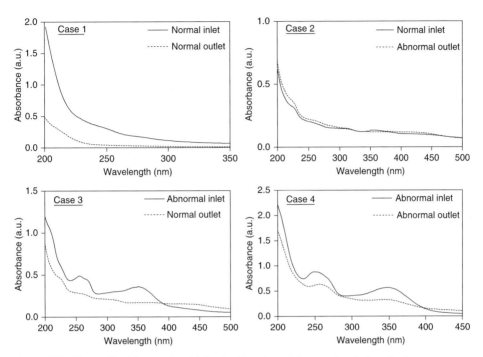

FIGURE 25. *Fate of incidents after biological treatment (example of chemical wastewater).*

TABLE 7. *COD values and removal yields*

Case	Observations about UV spectra	Inlet COD (mgO$_2$/L)	Outlet COD (mgO$_2$/L)	γ_r (%)
Case 1	*Si* reg/*So* reg	1100	180	84
Case 2	*Si* reg/*So* irreg	2080	1056	49
Case 3	*Si* irreg/*So* reg	3230	1755	45
Case 4	*Si* irreg/*So* irreg	1900	1080	43

method can be overcome by the use of complementary analysis (IR spectrophotometry, TOC measurement, etc.) in order to detect the responsible compounds.

Case 3 presents the situation where the accidental effluent has been managed, thanks to the photodegradability test. The characteristic shape of inlet UV spectrum is due to the presence of unknown absorbing compounds. Figure 26 presents the evolution of inlet UV spectrum according to irradiation time: after 12 min, it can be noticed that the UV spectrum is quite close to a regular spectrum. The medium yield can be imputed to the low activity of biomass not yet adapted to new compounds.

Case 4 describes the situation where UV spectra (*Si* and *So*) are irregular and have, moreover, a similar structured shape. The characteristic shape is still present in *So*, showing that the major pollutants are partially degraded.

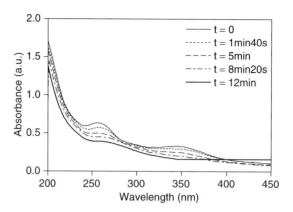

FIGURE 26. *Photodegradation test of inlet effluent (case 3): evolution of UV spectra according to irradiation time (pathlength: 2 mm, no dilution).*

Case 5 (not shown) would be characterised by irregular UV spectra of different shapes related to the potential presence of heterogeneous materials, for example, coming from the biomass death (in case of a biological treatment plant).

In order to prevent and better manage the unexpected situations (Cases 2, 4 and 5), UV spectra of accidental effluents can be stored in a data bank that can be systematically consulted for the admission of effluents in the wastewater treatment plant.

4.5. External waste management

A last application concerning the use of UV spectrophotometry for waste management is the quality control of external wastes, brought on a given centralised treatment plant. Wastes from septic tanks, industrial liquid wastes (high loaded wastewater or process water) or washing effluents (from tanks, for example) can be collected in small amounts by trucks, and brought to a centralised treatment plant.

These wastes of various origins (Table 8) are characterised by a huge qualitative and quantitative variability. However, all these wastes present more or less structured UV spectra as shown in Fig. 27.

TABLE 8. *Main pollutants associated to external wastes*

Industry or water type	Main pollutants
Perfumeries	Organic matter, solvents (alcohol, ketones, etc.)
Fine chemistry	Various chemical products
Electroplating	Surfactants, citric acid, various minerals, etc. ...
Leachates	Organic matter, surfactants, etc.
Water with hydrocarbons	Hydrocarbons, alcohols
Trucks and tanks washing water	Various chemical products
Chemical process water	Specific chemical products

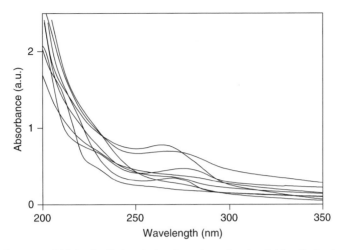

FIGURE 27. *Spectra of biologically treated external wastes (variable dilution).*

In order to control the quality of wastes brought to the treatment plant, a simple procedure can be used for the reception of trucks and the acceptance of liquid wastes. Based
on the acquisition of UV spectrum of the waste to be treated, a comparison with a reference spectrum shape (selected from a spectra library, characteristics of the different
types of wastes) is carried out (Fig. 28). In function of the comparison result (by using
one of the tools presented in Chapter 2), the waste is accepted or rejected.

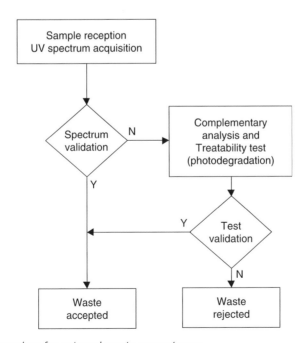

FIGURE 28. *Procedure for external waste acceptance.*

With respect to the previous procedure based on the measurement of classical parameters (COD, TOC, etc.), this new one leads to a more relevant quality control with a time gain leading to the increase of trucks travels.

5. ENVIRONMENTAL IMPACT

5.1. Discharge

UV spectrophotometry can also be used, such as for urban wastewater (see Chapter 8), for the study of the environmental impact of treated wastewater. Figure 29 displays the different spectra acquired for a dispersion study of treated industrial (petrochemistry) wastewater discharge in seawater. The set of spectra shows an isosbestic point due to the mixing of wastewater with seawater, the spectrum of which presents a sharp absorbance at 205 nm due to chloride (notice that the difference between spectra of nitrate and chloride solutions is chemometrically easy to exploit). After 100 m, the impact of the discharge is difficult to see.

5.2. Groundwater survey

A last application concerns the groundwater quality survey of an industrial site. Specific industries, such as, for example, refineries, petrochemical or chemical sites, are obliged to control the groundwater quality. For the purpose, several boreholes (wells) are used, from which groundwater is regularly sampled (for example, every month) for analysis. The use of UV spectrophotometry can give interesting qualitative (and quantitative) information, as can be seen in Figs. 30 and 31 [3].

In Fig. 30 are displayed the spectra of groundwater sampled from five wells of an industrial site (petrochemistry). The examination of spectra shape shows that one area (well 1) is highly contaminated with a major pollutant (not identified), presenting two peaks

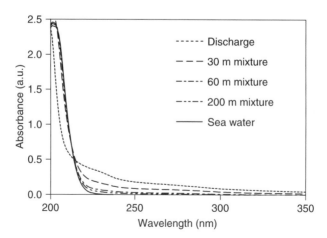

FIGURE 29. *UV spectra of industrially treated wastewater discharge and receiving medium (seawater).*

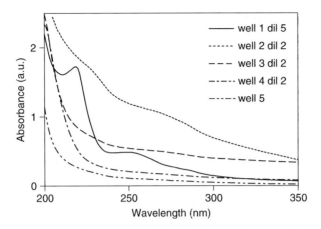

FIGURE 30. *Space variation of UV spectra of groundwater of a petrochemical site.*

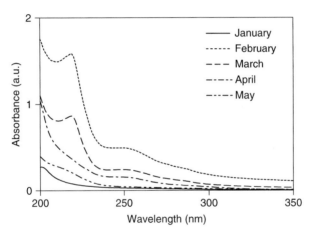

FIGURE 31. *Time variation of UV spectra of groundwater of a petrochemical site (well 1, Fig. 30).*

of absorption around 220 and 250 nm. Furthermore, wells 2 and 3 present a residual diffuse absorbance (above 300 nm), probably related to emulsified hydrocarbons.

The evolution with time of the groundwater quality of well 1 (Fig. 31) shows that the detected pollution has appeared in February and was still present (but diluted) in April.

References

1. D.L. Ford, J.M. Eller, E.F. Gloyna, *J. Wat. Pollut. Control Fed.*, 43 (1971) 1712.
2. O. Thomas, *Métrologie des Eaux Résiduaires*, Tec et Doc, Lavoisier, Paris, Liège, (1995).
3. H. El Khorassani, *Caractérisation d'Effluents Industriels par Spectrophotométrie Ultra-violette Appliquée à l'Industrie Pétrochimique*. PhD thesis, Université de Provence, France, (1998).

4. E. Baurès, *La Mesure Non Paramétrique, un Nouvel Outil pour l'Étude des Effluents Industriels: Application aux Eaux Résiduaires d'une Raffinerie*. PhD thesis, Université d'Aix-Marseille III, France, (2002).
5. F. Pouly, *La Spectrophotométrie UV: Une Approche Analytique Alternative aux Besoins du Raffinage*. PhD thesis, Université de Provence, France, (2001).
6. M-F. Pouet, E. Baurès, S. Vaillant, O. Thomas, *Appl. Spectrosc.* 58 (2004) 486.
7. O. Thomas, E. Baurès, M-F. Pouet, *Water Qual. Res. J.* Canada 40 (2005) 51.
8. O. Thomas, F. Théraulaz, V. Cerdà, D. Constant, P. Quevauviller, *Trends Anal. Chem.*, 16 (1997) 419.
9. F. Pouly, E. Touraud, C. Langellier, C. Busatto, O. Thomas, *Hydrocarbon Processing*, 1 (2001) 76.
10. S. Vaillant, M-F. Pouet, O. Thomas, *Talanta*, 50 (1999) 729.
11. S. Vaillant, *La Matière Organique des Eaux Résiduaires Urbaines: Caractérisation et Évolution*. PhD thesis, Université de Pau et des Pays de l'Adour, France, (2000).
12. S. Bayle, N. Azéma, C. Behro, M-F. Pouet, J-M. Lopez-Cuesta, O. Thomas, *Colloids and Surfaces A: Physicochem. Eng. Aspects* 262 (2005) 242.
13. C. Behro, *Caractérisation de Fractions Hétérogènes par Méthodes Optiques au cours de Procédés de Fabrication. Application aux Suspensions Papetières et Vinicoles*. PhD thesis, Université de Pau et des Pays de l'Adour, France, (2003).
14. B. Roig, C. Gonzalez, O. Thomas, *Anal. Chim. Acta*, 389 (1999) 267.
15. B. Roig, C. Gonzalez, O. Thomas, *Talanta*, 50 (1999) 751.
16. F. Pouly, E. Touraud, J.F. Buisson, O. Thomas, *Talanta*, 50 (1999) 737.
17. B. Roig, E. Chalmin, E. Touraud, O. Thomas, *Talanta*, 56 (2002) 585.
18. H. El Khorassani, G. Besson, O. Thomas, *Quimica Analitica*, 16 (1997) 239.
19. F. Theraulaz, L. Djellal, O. Thomas, *Tenside Surf. Det.*, 33 (1996) 447.
20. B. Roig, C. Gonzalez, O. Thomas, *Spectrochimica Acta part A*, 59 (2003) 303.
21. H. El Khorassani, P. Trebuchon, H. Bitar, O. Thomas, *Wat. Sci. Tech.*, 39 (1999) 77.
22. Y. Coque, *Proposition d'Outils d'Optimisation de Procédé d'Oxydation Avancé (POA) par UV/H_2O_2*. PhD thesis, Université de Pau et des Pays de l'Adour, France, (2002).
23. O. Thomas, F. Théraulaz, *Trends Anal. Chem.*, 13 (1994) 344.
24. L. Djellal, F. Theraulaz, O. Thomas, *Tenside Surf. Det.*, 34 (1997) 316.
25. E. Baurès, C. Behro, M-F. Pouet, O. Thomas, *Wat. Sci. Tech.*, 49 (2004) 486.
26. L. Castillo, H. El Khorassani, P. Trebuchon, O. Thomas, *Wat. Sci. Tech.*, 39 (1999) 17
27. C. Marty-Muret, *Estimation de la Traitabilité Potentielle d'Eaux Résiduaires Industrielles par de Nouveaux Paramètres de Caractérisation*. PhD thesis, INSA Lyon, France, (2001).
28. C. Muret, M-F. Pouet, E. Touraud, O. Thomas, *Water Sci. Tech.*, 42 (2000) 47.
29. H. El Khorassani, P. Trebuchon, H. Bitar, O. Thomas, *Wat. Sci. Tech.*, 42 (2000) 15.

UV-Visible Spectrophotometry of Water and Wastewater
O. Thomas and C. Burgess (Eds.)

243

CHAPTER 10

Leachates and Organic Extracts from Solids

E. Touraud[a], J. Roussy[a], M. Domeizel[b], G. Junqua[c], O. Thomas[c]

[a]Laboratoire Génie de l'Environnement Industriel, Ecole des Mines d'Alès, 6 Avenue de Clavières,
30319 Alès Cedex, France; [b]Laboratoire Chimie et Environnement, Université de Provence,
3 place V. Hugo, 13331 Marseille Cedex, France; [c]Observatoire de l'Environnement et du
Développement Durable, Université de Sherbrooke, Sherbrooke, Québec, J1K 2R1, Canada

1. INTRODUCTION

Leachates are produced by percolation of rain water through a solid matrix, such as solid wastes from urban or industrial landfills, or polluted soils, for example (Fig. 1). Leachates are highly polluted solutions (or suspensions) characterised by a high salinity and organic content. Consequently, they are rather complex to study or analyse, as huge interferences may occur and affect results.

Regarding leachate analysis, UV spectrophotometry can be useful for a fast characterisation or the study of landfill evolution. Depending on the nature of organic components, aqueous solutions can limit the interest of the approach. In this case, an extraction step of the solid matrix with an organic solution can be necessary in order to have more specific information.

In this chapter, some applications dealing with landfill leachates, contaminated soils and solid waste composts are presented. A complementary study concerning natural sediments is presented at the end of the chapter. If needed, extraction procedures must be chosen as simple as possible, but other methods or techniques could be used for the purpose. In this field in particular, further developments or applications are easy to imagine.

2. LANDFILL LEACHATES

Landfill leachates are generally considered to be highly polluted media, containing various organic compounds refractory to biodegradation. In fact, their composition varies, on one hand, with the nature of disposed solid wastes, and, on the other hand, with the landfill storage duration and the treatment process, if any. Organic matter of landfill leachates can be characterised by the determination of various parameters. Specific organic compounds are detected by chromatographic techniques [1,2]. Global parameters such as chemical oxygen demand (COD), biological oxygen demand (BOD), dissolved organic carbon (DOC), nitrogen forms, chemical oxygen demand and optical methods, and particularly UV spectrophotometry, are also used [2]. The following examples deal with the characterisation of leachates, with respect to their origin, and present some treatment experiments.

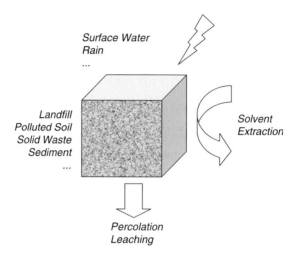

Surface Water
Rain
...

Landfill
Polluted Soil
Solid Waste
Sediment
...

Solvent
Extraction

Percolation
Leaching

FIGURE 1. *Transfer mechanisms from solid to liquid phase.*

2.1. Leachate characterisation

2.1.1. Direct examination of UV spectra

UV-visible spectra of leachates from four municipal solid waste landfills, Augsburg, Grospierres, Munich and Saint-Brès are presented in Fig. 2. All these samples are diluted 20 times, except Saint Brès (40 times). The direct examination shows, for all leachate samples, a decreasing monotonous spectrum between 200 and 400 nm. Nevertheless, some individual particularities can be noticed.

Augsburg and Munich leachates are characterised by a rather low absorbance ratio between the beginning and the middle of the spectrum (300 nm). At the same dilution, the spectrum of Grospierres leachate shows a higher absorbance on the studied range,

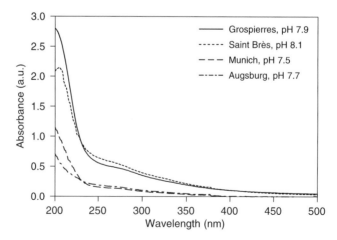

FIGURE 2. *UV spectra of urban landfill leachates (diluted 50).*

TABLE 1. *Aggregate parameters of the studied landfill leachates*

Leachate	COD (mgO$_2$/L)	DOC (mgC/L)	Conductivity (mS·cm^{-1}) [Cl$^-$ (mg/L)]
Augsburg	550	230	4.8 [430]
Grospierres	930	160	3.1 [540]
Munich	690	150	9.0 [2550]
Saint-Brès	1560	670	6.4 [1100]

suggesting a higher organic load. The same hypothesis can be formulated for Saint-Brès leachate, which is more diluted. A shoulder is noted around 270–280 nm, more important for the Grospierres and Saint-Brès samples. These observations are confirmed by the values of other parameters such as COD and DOC (Table 1).

The spectrum of Munich leachate is more concave with no marked accident, and its absorbance decreases rapidly between 200 and 230 nm, due to a more important salinity, especially chloride ion. This spectrum also gives an indication of the composition of the leachate. Indeed, simple organic molecules that result from the initial degradation process absorb at the beginning of the UV wavelength range, and few humic-like substances are present (low shoulder at 270 nm).

2.1.2. pH effect

The studied leachates have a pH close to 7.5–8.5. Acidification has been then made by the addition of sulphuric acid until pH 2. Munich and Augsburg leachates show the same behaviour and are globally not affected by the pH change, proving the low concentration of humic-like substances (Fig. 3).

Augsburg and Munich leachates are characterised by an important level of inorganic carbon, respectively, 78% and 71% of total carbon (Table 2). Indeed, as a small difference exists at the beginning of their UV spectra, it can be assumed that inorganic carbon is eliminated by acidification, and the decrease in absorbance can be due to the

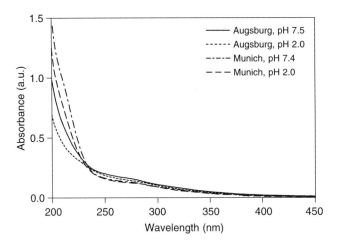

FIGURE 3. *pH effect on Augsburg and Munich leachates (diluted 5).*

TABLE 2. *Carbon concentration (organic, DOC, and inorganic, IC) for Augsburg and Munich leachates*

Leachate	pH	DOC (mgC/L)	IC (mgC/L)
Augsburg	7.7	160	560
	2.0	160	<1
Munich	7.5	140	350
	2.0	140	<1

disappearance of hydrogenocarbonate ions, which slightly absorbs in this wavelength window.

On the contrary, Grospierres and Saint-Brès leachates are sensitive to pH (Fig. 4). After acidification, the shape of their UV spectra is the same, but the absorbance decreases all along the wavelength range; the organic load is lowered, as shown by COD and DOC values before and after acidification (Table 3). In the same way, the shoulder around 270–280 nm is less important after treatment. This phenomenon is supported by the observation of a precipitate after acidification, probably humic-like substances.

2.2. Leachate treatment

UV spectrophotometry also provides information about the evolution of leachates during classical treatments. Some treatment trials have been made on the previous landfill

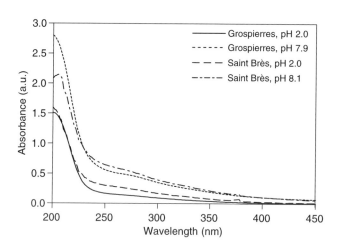

FIGURE 4. *pH effect on Grospierres and Saint-Brès leachates (diluted 5).*

TABLE 3. *pH effect on the organic load of Grospierres and Saint-Brès leachates*

	Grospierres	Saint-Brès
COD (mgO$_2$/L) before acidification	930	1560
COD (mgO$_2$/L) after acidification	520	1040
COD reduction (%)	44	33

leachates. Coagulation–flocculation tests with $FeCl_3$ (500 mg/L) and UV photo-oxidation coupled or not with a chemical oxidising agent (H_2O_2, 0.1 mole/L) have been tested. UV spectra before and after treatment are presented.

2.2.1. Coagulation–flocculation with $FeCl_3$

Coagulation–flocculation with $FeCl_3$ only enables the elimination of a rather small part of the organic load. This part is nevertheless more important in acidic conditions as shown for Grospierres leachate in Fig. 5. COD data before and after treatment are given in Table 4 for Grospierres leachate. The results show that a pH decrease to 5, in the presence of $FeCl_3$, leads to the removal of about 30% of the COD, probably explained by the precipitation of humic-like substances.

2.2.2. Photo-oxidation

The efficiency of photo-oxidation treatment on leachates (at pH 2) has been studied with UV spectrophotometry. Different attempts have been made in presence of an oxidising agent, of variable concentration (Table 5). Figures 6 and 7 present, respectively, the results for Augsburg and Saint Bres leachates, for an irradiation time of 15 min (and H_2O_2 0.1 mole/L). A real efficiency is observed both from the evolution of TOC (or DOC) and UV spectra.

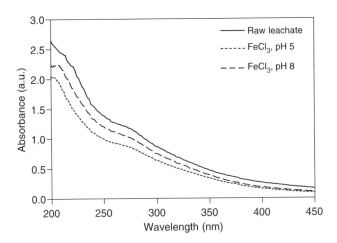

FIGURE 5. *Effect of coagulation–flocculation on Grospierres leachate ($FeCl_3 = 500$ mg/L).*

TABLE 4. *COD data for Grospierres leachate treatment (coagulation–flocculation)*

	pH 5	pH 8
COD (mgO_2/L) before treatment	930	930
COD (mgO_2/L) after treatment	660	800
COD reduction (%)	29	14

T_ABLE 5. *Effect of H_2O_2 concentration on TOC reduction*

H_2O_2 concentration (mole/L)	TOC reduction (%) Grospierres leachate	TOC reduction (%) Augsburg leachate
0.01	10	—
0.025	20	75
0.05	62	87
0.1	86	91

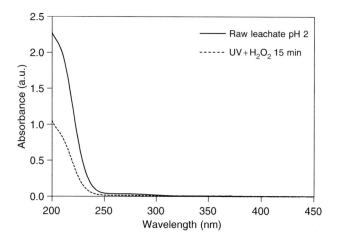

F_IGURE 6. *UV spectra of from Augsburg leachate photo-oxidation (dilution 5).*

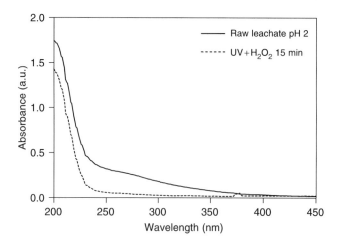

F_IGURE 7. *UV spectra of Saint-Brès leachate photo-oxidation (dilution 40).*

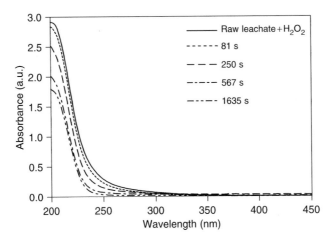

FIGURE 8. *UV spectra of Grospierres leachate photo-oxidation according to the irradiation time.*

The specific shoulder around 270–280 nm disappears after treatment, which is a sign of organic load degradation. For the two studied leachates, significant absorbance values are obtained at the beginning of the UV range (200–240 nm), where degradation products, especially organic acids or nitrate/nitrite ions show specific absorption (around 220 nm).

Finally, the influence of the irradiation time is reported in Fig. 8 for Grospierres leachate (pH 5). The decrease in absorbance is more important when the irradiation time increases, as expected. This phenomenon is correlated to the DOC reduction (DOCf/DOCi), which has been measured during the photo-oxidation tests (Table 6).

In conclusion, the direct examination of UV spectra reveals some differences according to the organic load and to the management of urban solid waste landfill. For instance, Fig. 1 shows quite different UV spectra for Grospierres and Munich, which have nearly the same DOC value. This can be explained by the variability of the nature of soluble organic matter that can be affected by the origin of the wastes (rural versus urban). On the other hand, higher organic loads could be expected for urban landfills (Augsburg and Munich). UV spectra do not confirm this hypothesis, and one may suppose that these leachates have been pretreated (e.g. coagulation–flocculation). Indeed, no precipitate was observed either by acidification at pH 2 or by addition of $FeCl_3$.

TABLE 6. *DOC reduction according to irradiation time (Grospierre leachate)*

Irradiation times	DOCf/DOCi
0	1
81	0.75
250	0.46
567	0.29
1635	0.10

3. POLLUTED SOILS

Polluted soils can often be considered as ancient raw industrial landfills. They are characterised by high concentrations of specific organic (or mineral) compounds related to the
processes that were used on the site. Pollutants to be determined are varied, generally
toxic and persistent, since they can remain in the soil for a long period after the end of the
industrial activity. Among the main priority compounds, polycyclic aromatic hydrocarbons
(PAHs) and petroleum hydrocarbons are often encountered in ancient coking plants, gas
works or refinery sites. Since such pollutants have to be considered as potential pollutants for groundwater, it is interesting to show how UV-visible spectrophotometry can be
applied for their study. Considering their characteristics and the matrix complexity, an
extraction step is necessary before spectrum acquisition.

3.1. Polluted soils characterisation

The extraction step can be made either in aqueous or organic medium. Before extraction, the soil sample is pretreated (drying, grinding and sieving) according to the first
steps of the classical procedure [3]. The solid/liquid ratio used for the extraction is 1 g
soil/10 mL solvent.

3.1.1. Polycyclic aromatic hydrocarbons

Two PAH-contaminated soils of different origins have been studied. Soil A (sandy soil)
comes from an ancient coking plant, and soil B (clay soil), from a rather recent creosote
production site. Spectra have been acquired after mechanical agitation with deionised
water for 1 and 24 h (Fig. 9). Direct examination of aqueous sample spectra leads to the
following observations

- Water-soluble matter and UV absorbance increase with extraction time, but the shape
  of UV spectra is not modified for each studied soil.

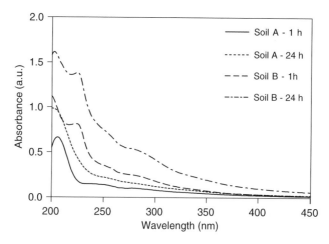

FIGURE 9. UV spectra of PAH-contaminated-soil aqueous leachates (dilution 10).

- On the opposite, UV spectra shapes are quite different according to the nature and age of pollution. Soil A shows concave spectra with an absorbance that decreases rapidly beyond 210 nm, showing that the degradation process is advanced. Nevertheless, a small shoulder is still noted, around 250 nm.
- UV spectra of soil B present a marked shoulder at 225 nm and a significant absorbance up to 350 nm. Small shoulders are observed around 250 nm and 270–280 nm. These particularities are characteristic of industrial pollutants such as phenols, which can be found in PAH-contaminated soils and compounds belonging to the PAH family. Moreover, the residual pollution is more important in soil B.

UV-visible spectrophotometry enables giving indications about the pollution maturation level in contaminated soils. Stabilised leachates from old contaminated soils are characterised by a monotonous decreasing spectrum (soil A), while younger ones show a specific spectrum where additional compounds are responsible for visible accidents (soil B).

In order to avoid huge interferences from the matrix (humic-like substances), extraction by sonication using acetonitrile has been tested on the two previous contaminated soils. After 1 h of sonication, UV spectra of the supernatant have been acquired (Fig. 10).

From a qualitative point of view, direct examination of UV spectra shows approximately the same shape for the two soils, with many visible accidents due to the presence of specific compounds such as PAHs. Moreover, because of the dilution factor, it can be assumed that acetonitrile extraction is more efficient than water extraction, especially for nonpolar compounds such as PAHs. As expected, the two specific peaks located at 254 and 288 nm can be noticed on the two UV spectra.

From a quantitative point of view, the global PAH concentration can be estimated with the use of PAH index previously defined in Chapter 4. The value of the ratio $A_{254\ nm}/A_{288\ nm}$ gives information on the distribution of PAH in the contaminated soils, namely that as higher is this ratio, and higher is the proportion of light PAHs. Table 7 collects the data related to the two studied contaminated soils.

For field application, the sonication step for acetonitrile extraction can be replaced by 10 min of manual extraction with hand. This procedure has been compared to sonication and led to similar results [3].

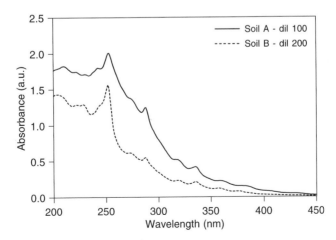

FIGURE 10. *UV spectra of PAH-contaminated soils acetonitrile extracts.*

TABLE 7. *Data related to PAH-contaminated soils (A and B)*

Sample	PAH UV (g/kg)	$A_{254\ nm}/A_{288\ nm}$
Soil A	4.3	1.61
Soil B	6.7	2.87

3.1.2. Petroleum hydrocarbons

Figure 11 presents spectra of water extract (dilution 10) and acetonitrile extract (dilution 250) of a soil contaminated by petroleum hydrocarbons. As for PAH extraction, the use of an organic solvent is more efficient. The corresponding spectrum is more structured, showing two specific peaks, located at 228 and 256 nm. Most unsaturated hydrocarbons, especially alkylated derivatives of benzene, absorb in this region. Nonaromatic hydrocarbons absorb mainly in vacuum UV.

The UV-visible spectrum of aqueous extract is quite monotonous. A marked shoulder can be noted around 210 nm and beyond, the absorbance decreases quickly until 250 nm and then rather smoothly until 350 nm. The specific and strong absorbance at 210 nm can be imputed to the degradation products of petroleum hydrocarbons, especially carboxylic acids.

3.2. Treatment of polluted soils

3.2.1. Polycyclic aromatic hydrocarbons (PAH)

The monitoring of a biological treatment carried out in laboratory has been done for two PAH-contaminated soils, soil 1 mainly polluted by heavy PAHs, and soil 2 mainly contaminated by light PAHs. UV spectra according to treatment time are given in Fig. 12, for the two experiments.

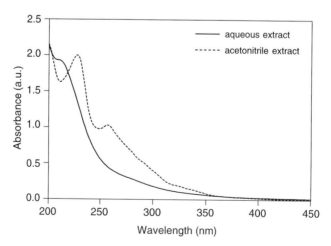

FIGURE 11. *UV spectra of aqueous and acetonitrile extracts of soil contaminated by petroleum hydrocarbons (dilution 10 for aqueous extract, 250 for acetonitrile extract).*

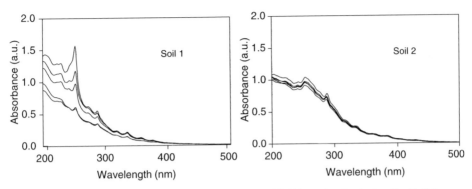

FIGURE 12. *UV monitoring of biological treatment of PAH-contaminated soils. Soil 1, samples at 0, 35, 50, 50 and 220 days, and Soil 2, samples at 0, 40, 80, 100 and 130 days (by decreasing order of absorbance at 200 nm).*

Table 8 presents the evolution of, on one hand, the global PAH concentration (measured by HPLC and estimated by UV) and, on the other hand, the ratio, $A_{254\ nm}/A_{288\ nm}$.

No significant difference exists between HPLC measurement and UV estimation. The yield of decontamination is roughly of 75% for soil 1, after 110 days of treatment. For soil 2, it aims for only 20% after 130 days of treatment. These results are in agreement with those of the literature. Heavy PAHs are less sensitive to microbiological degradation during biological treatment, which takes place into the soil under natural conditions [4].

The ratio $A_{254\ nm}/A_{288\ nm}$ decreases regularly with treatment time from 2.9 to 1.7 for soil 1 mainly polluted by light PAHs, but remains constant and equal to 1.3 for soil 2. This ratio, the value of which is connected to the proportion of light PAHs in the contaminated soil, reflects the degradation degree of PAH during biological treatment (Fig. 13). It may be considered as a soil evolution index and allows estimating the potential biological treatability of a PAH-contaminated soil.

In conclusion, PAH and soil evolution indexes are simple and rapid tools for the characterisation of PAH-contaminated soils in terms of level and repartition of pollution and for the prediction of their potential biotreatability. From an environmental point of view, they permit pointing out sensitive zones and defining priorities in terms of decontamination.

TABLE 8. *Data related to monitoring of PAH-contaminated soil (1 and 2) treatment*

Sample	PAH HPLC (g/kg)	PAH UV (g/kg)	$A_{254\ nm}/A_{288\ nm}$
Soil 1/0 day	8.2	6.7	2.89
Soil 1/35 days	4.2	4.9	2.29
Soil 1/50 days	3.7	3.9	2.12
Soil 1/110 days	2.2	2.2	1.76
Soil 1/220 days	2.0	2.2	1.67
Soil 2/0 day	2.9	2.3	1.32
Soil 2/40 days	2.9	2.0	1.33
Soil 2/80 days	2.8	1.9	1.35
Soil 2/100 days	2.8	1.9	1.35
Soil 2/130 days	2.3	1.8	1.35

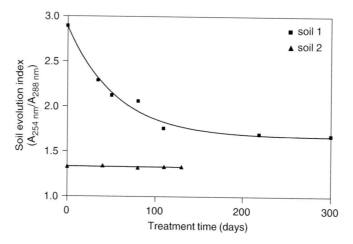

FIGURE 13. *Evolution of soil evolution index during biological treatment.*

For field analysis, a kit, using a field UV spectrophotometer, has been developed and gives the two index values directly [5].

3.2.2. Petroleum hydrocarbons

Figure 14 presents UV-visible spectra of the acetonitrile extracts of a petroleum-hydrocarbon-contaminated soil during a biological treatment on site (biopiles). The treatment time increases from sample 1 to sample 4 and, as far as the UV spectrum is less structured, the composting process is more advanced. In the same time, the pollution level decreases as the absorbance values all over the UV range, and particularly at 228 and 256 nm, decrease.

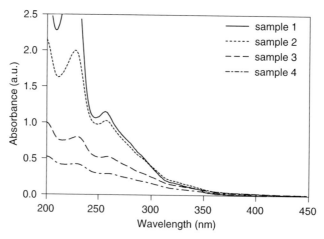

FIGURE 14. *UV monitoring of a petroleum hydrocarbon soil biological treatment.*

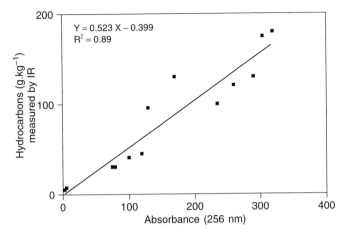

FIGURE 15. *Correlation between IR and UV measurements of petroleum hydrocarbons.*

Hydrocarbon concentration has been measured by IR spectrophotometry, after a carbon tetrachloride extraction step [6]. The UV estimation (256 nm absorptiometry) has been tested compared to IR data for the 15 studied polluted soils (Fig. 15). The absorbance ratio $A_{228\ nm}/A_{256\ nm}$ can reflect, during a biological treatment, the composting degree of soils contaminated by petroleum hydrocarbons, as for PAH pollution. Quantitative data are collected in Table 9.

As can be noted, the composting process starts ($A_{228\ nm}/A_{256\ nm} > 1.9$) at the beginning of the biological treatment on site (sample 1). During the next steps (samples 2 and 3), the composting is efficient ($1.9 < A_{228\ nm}/A_{256\ nm} < 1.5$). A value of $A_{228\ nm}/A_{256\ nm}$ lower than 1.5 for sample 4 indicates that the final composting phase is reached, i.e. that decontaminated soils can be reused.

In conclusion, simple UV indexes presented in this chapter enable a rapid diagnosis of hydrocarbonated pollution in terms of location, distribution and aging. They can also be useful in the monitoring of natural attenuation.

4. SOLID WASTES TREATMENT BY COMPOSTING

Solid wastes treatment is nowadays very important because, on the one hand, their quantity increases regularly and, on the other hand, their storage in landfills tends to be forbidden. This is particularly true for urban solid wastes. In this frame, the biological treatment of solid wastes, or composting, often associated with biological sludges from

TABLE 9. *Monitoring of the composting of petroleum-hydrocarbon-polluted soils*

Sample	HC IR (%)	HC UV est. (%)	$A_{228\ nm}/A_{256\ nm}$
1	13	15	2.0
2	12	13.5	1.9
3	9.5	7	1.7
4	3	4	1.5

wastewater treatment plant or with green wastes, can be an economical solution. As for other processes, the quality control of the mixture, the compost, has to be carefully checked during the treatment, particularly because the destination of the final mature compost (in fact, mineralised) may be for an agricultural use.

4.1. Characterisation of solid wastes

The aim of solid wastes treatment is both to reduce their size and to stabilise their organic content. In order to obtain qualitative and quantitative information about humification of anthropogenic organic matter during the treatment process or waste valorisation, a simple analytical method has been developed [7].

Many tests and criteria have been proposed in order to evaluate the maturity level of compost, but they are generally time consuming and not easy to run. The humification process and organic matter evolution can be followed with two approaches:

- The first is based on the structural evolution of the organic matrix with the character-isation of functional or structural groups. Fluorescence spectroscopy, infrared spec-troscopy or RMN measurement have been developed to follow humic-like-substances in soil, and to study urban compost [8–13].
- The second approach consists of the study of both humic and fulvic fractions evo-lution. This way seems to be more operational. The evolution of the humic nature of the organic matter during composting or after mixture with soil can be estimated by one of the following methods: C/N ratio [14–16], humification index [17–19] or biodegradability index [11,14]. Another parameter, based on the use of two wave-lengths (465 and 665 nm), can also be used for the characterisation of the evolution of humic-like substances [20,21].

The use of UV spectrophotometry has been studied in order to evaluate the potentiality of this technique for the proposal of an index of maturity. Figure 16 shows the aque-ous leachates of composts with a great difference between fresh and treated compost.

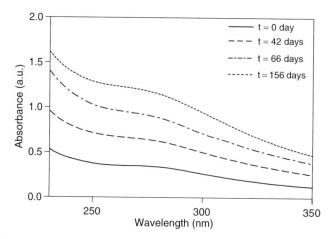

FIGURE 16. UV spectra evolution according to composting time.

This last seems to contain more soluble and absorbing compounds than the fresh one, and the increase of absorbance is continuous during the treatment, with an enlargement of a shoulder at around 260–270 nm [7,20].

As the UV spectrum shape is relatively featureless with a large shoulder at 260–280 nm, a complementary procedure is used for a fine characterisation of organic matter extracted by a basic medium ($K_4P_2O_7$). This method, already used in a previous work [22], is based on the separation of the extract by low-pressure gel chromatography with a study of the fractions by UV-spectrophotometry. The selected gel (Sephadex G75) allows the separation of macromolecules of apparent molecular weight (MW) between 3000 and 70,000 Da. During the elution process, the molecules with higher MW are eluted first, the smaller ones being retained in the gel according to their MW (Fig. 17). The experiment has been carried out on leachates corresponding to the beginning and end of the compost process of an urban waste (156 days of composting). The examination of chromatograms and the corresponding UV spectra shapes show the organic matter evolution during composting.

The chromatogram corresponding to the fresh compost is characterised by three peaks of absorbance (followed at 270 nm), the first being more intense. The UV spectra of

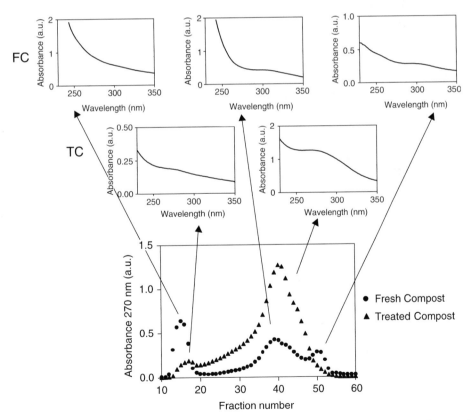

FIGURE 17. *Chromatograms (and UV spectra of fractions) of extracts of fresh and treated composts (FC and TC).*

the related fractions (15, 39 and 52) are different, but without specific feature. For the matured compost, the chromatogram is simpler with only two peaks, the second peak corresponding to high values of absorbance at 270 nm. The related fractions (17 and 42) seem to correspond to the two first fractions of the fresh compost but their UV spectra are different, showing a clear shoulder at 260–270 nm. Assuming that the two first fractions of each leachate correspond roughly to the same groups of compounds, the evolution during composting tends to lower their MW with degradation process. More precisely, compounds with apparent MW >70,000 Da (fraction 15 of fresh compost) are degraded into smaller molecules, as it can be seen with the strong decrease of the first peak intensity of the matured compost.

At the same time, compounds formed during the composting process seem to be characterised not only by lower MW than the initial ones, but also by the presence of aromatic structures explaining the shoulder of the corresponding UV spectrum (fraction 42 of the final compost). For example, phenolic compounds generally considered to be the building blocks of humic substances has been identified [19,23] with other functional groups (carboxylic and hydroxylic).

Thus, during the composting process, mineralisation is a competitive reaction with respect to humification, characterised by a high aromaticity level. The latter observation demonstrates the increase of complexity of organic matter with increasing maturity level.

Faced with this complexity, the proposal of a maturity index from UV spectrophotometric data supposes considering the evolution of the whole spectrum. The first step is the identification of the particular experimental spectra that permit explaining any other spectrum as a linear combination of these reference spectra [24]. This procedure is the semi-deterministic one, presented in Chapter 2 and already applied in the previous applications. These spectra are automatically selected from the set of a great number of spectra of raw extracts of composts (including humic and fulvic acids) or related to a gel chromatographic fraction. Four reference spectra have been selected from the spectra bank, corresponding respectively to the spectra of (Fig. 18):

- Humic acid extracted from a 6-month-old compost (first reference spectrum named ref 1 in Fig. 18),
- Fraction 45 obtained by G75 gel chromatography separation of the $K_4P_2O_7$ extract from a 12-month-old compost (ref 2),
- Fraction 41 obtained by G75 gel chromatography separation of the $K_4P_2O_7$ extract from a 3-month-old compost (ref 3),
- Humic acid extracted from 8-day-old compost (ref 4).

Notice that the automatic procedure, based on the "rank" matricial calculation [24], has chosen the most independent spectra related to some characteristics of the fresh and mature composts.

The second step of the study of a maturity index is the deconvolution of spectra of $K_4P_2O_7$ extracts obtained from composts of various ages, with the basis of reference spectra. The coefficient contribution variation, related to the reference spectra, is followed with time. The general tendency is an increase with time of the coefficient of spectra 1 and 2 (called C1 and C2, respectively) and a decrease of the coefficient of spectra 3 and 4 (called C3 and C4, respectively).

The composition of the basis of reference spectra and the general evolution of the coefficients are closely linked to the various states of the evolution of organic matter during composting. A maturity index can thus be proposed from the values of these coefficients.

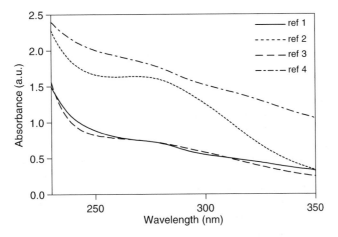

FIGURE 18. *Reference spectra for the deconvolution of compost extracts.*

A study performed over two periods of 12 months composting has shown that the C1/C3 ratio increases during composting time and reaches a stable level after a maturation period of nearly 6 months for the two composts, A and B (Fig. 19). During the same time the value of the ratio (final *C/N*)/(initial *C/N*) reaches around 0.5, which is within the range currently accepted for compost more than 3 months old, and the ratio (humic acid)/(total humic substances) increases from 0.04 to 0.6 [7]. These results lead to the proposal the C1/C3 ratio as an index of maturity. This index is easy to run for the monitoring of a composting process. Its determination needs some glassware for the $K_4P_2O_7$ extraction and a UV spectrophotometer with multicomponent software.

UV spectroscopy is thus an interesting analytical tool to study and monitor organic matter humification. In spite of some necessary validation, this method defines a new, easy-to-run maturity index of compost. Of course, this new method can also be used

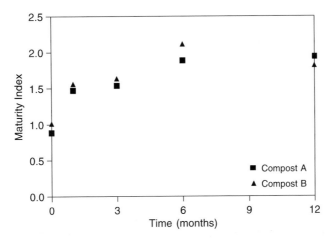

FIGURE 19. *Evolution of the maturity index of two composts (ratio of C1/C3; see text).*

to evaluate natural organic matter of soil and sediments as shown on the following application.

5.   NATURAL SOILS AND SEDIMENTS

The characterisation of natural soils and sediments with UV spectrophotometry applied after leaching tests is possible [25,26]. UV spectrophotometry can also be used to estimate nitrates, DOC and DOC fractions after XAD fractionation in forest-floor leachates [27] or in agricultural soil water extracts [28]. The exploitation of spectra of aqueous and alkaline extracts allows determining oxidation and humification indexes [26], even if the corresponding UV spectra present different featureless and smooth shapes. Moreover, as some organic compounds of soils and sediments are not soluble in water, a complementary organic extraction step is carried out. The choice of acetonitrile, as for PAH and hydrocarbon characterisation, has been made because of the necessity of using a non-UV-absorbing solvent.

5.1. UV characterisation

UV spectra of acetonitrile extracts for different humus and sediments are presented in Figs. 20 and 21 [25].

5.1.1. Humus

The direct examination of the UV spectra of different types of humus (Fig. 20) shows a rather strong absorption at the beginning of the spectrum, which decreases rapidly over 220 nm. A clear shoulder at 274 nm can be noticed for the leafy humus. This shoulder is not so marked in the case of pine humus, for which a little shoulder appears

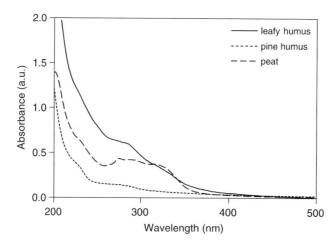

FIGURE 20. *UV-visible spectra of different types of humus (acetonitrile extracts, dilution 5).*

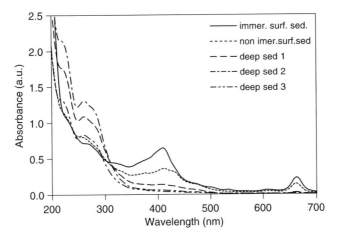

FIGURE 21. *UV-visible spectra of acetonitrile extracts of surface and deep sediments (no dilution for surface sediments, dilution 1/3 for deep sediments).*

at around 230 nm. Because its UV spectrum is not structured, it can be supposed that the degree of maturation is more advanced for the pine humus. The spectrum of peat presents some differences, corresponding to other natural conditions. A rather large shoulder between 250 and 350 nm suggests the presence of condensed organic matter, and the specific absorption at 210 nm (presence of nitrates) indicates that the mineralisation process is going on. These observations lead to the assumption that a typology of humus that takes into account the nature of vegetation and natural conditions, and based on UV-Visible spectra, could be proposed.

5.1.2. *Sediments*

Figure 21 presents UV-visible spectra of two types of sediments, surface sediments (immersed or not) and deeper sediments (10–15 cm of depth). The colour of the former ones is brown, and it is black for the latter ones. This difference can be noticed in the visible range of the spectra: a clear peak at 664 nm, specific of chlorophyll, can be noticed for surface sediments, which disappears for deeper sediments. This decrease of absorbance in the visible range goes with an increase of absorbance and a specific pattern of the spectrum in the UV region (200–300 nm), where degradation products may absorb. It may be supposed that, in this case, the process of decomposition of the organic matter is more advanced. This phenomenon is especially clear for deep sediments 1 and 2.

Figure 22 shows UV-visible spectra of interstitial water (raw and filtered) for deeper sediments 2 and 3. As expected, they are rather monotonous. For raw samples, the residual absorbance remains over 250 nm due to suspended matter. UV-visible spectra of filtered samples allow showing the absorption of organic matter characterised by a little shoulder at around 280 nm.

Fresh leaves have been extracted with acetonitrile in order to confirm the presence of chlorophyll in organic extracts of surface sediments (extract 2). A comparison has been made with an extraction with acetone (extract 1), according to standard methods [29].

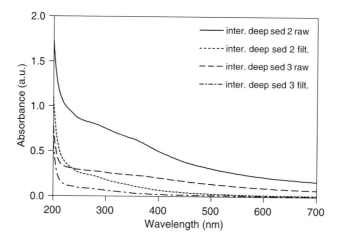

FIGURE 22. *UV-visible spectra of interstice waters before and after filtration (no dilution).*

UV-visible spectra have been acquired and seem very close (Fig. 23). The main difference is observed between 400 and 420 nm, where acetone still absorbs slightly. The absorption in the visible region observed previously (Fig. 21) is due to this pigment.

5.2. Application: evolution of sediments in wetlands

The quality of a peatland and its environment has been studied [25]. The Grand Mare (sampling stations 4 and 5) is a peatland of 47 ha, located in a wetland of the Seine estuary, the Vernier Marsh. Since it is the lowest part of the marsh, all annexes converge in it: le Ruel (sampling station 1), Petite Mare (sampling station 2) and la Crevasse

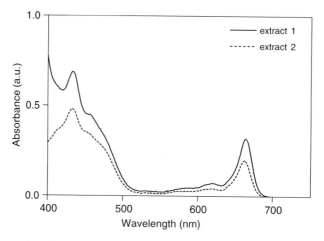

FIGURE 23. *Extraction of chlorophyll from fresh leaves with acetone (1) and acetonitrile (2).*

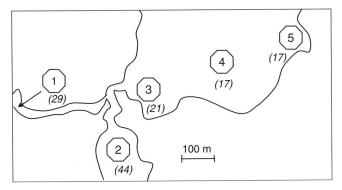

FIGURE 24. *Presentation of sampling stations in the Grand Mare and annexes (C/N ratio of the sediments).*

(sampling station 3). The Grand Mare is also connected with the Seine River, by a channel located in the north of the pond. Five sampling stations (Fig. 24) have been chosen in relation to their organic carbon concentration.

A procedure of acetonitrile extraction has been carried out on sediments. Figure 25 presents the UV spectra of the extracts. The shape of UV spectra is structured and characterised by the presence of a first peak at 275 nm and a second one at 410 nm. One can also note a first shoulder at 265 nm, and a second one at 315 nm.

The differences between samples of annexes (1, 2 and 3) and samples of the pond (4, 5) are the crests of the peaks. An evolution or humification index could be the ratio of the absorbance measured at these two wavelengths. The study of the C/N ratio evolution shows a gradient of mineralisation from the annexes to the Grand Mare.

Figure 26 shows the UV spectra of water from the same points. It is clear that stations 1, 2 and 3 are characterised by a high concentration of organic matter. A shoulder is observed at 270 nm, characteristic of aromatic structures such as humic-like

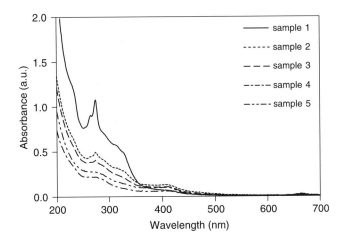

FIGURE 25. *Spectra of extracts.*

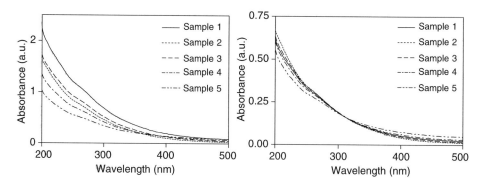

FIGURE 26. *UV spectra of water samples, diluted 5 times (left) and normalised spectra (right).*

substances (HLS), or precursors of HLS, potentially present in peatland. The dark colour of the samples can also be explained by this type of compounds (humified organic matter). In Grand Mare, the organic matter concentration is lower, and UV spectra are also less structured. This can be explained both by citing the reason of a dilution effect (inlet from the Seine River) and by citing a lesser input of fresh matter as the reason. Indeed, the stations 1, 2 and 4 are characterised by important vegetation (mainly trees such as birch) present on the banks, while Grand Mare is surrounded with reed bed.

The set of normalised spectra has an isosbestic point, illustrating the evolution of organic matter from simple molecules, giving a steeper slope at the beginning of the spectrum (samples 1 and 2), to complex molecules, smoothing the UV spectrum especially after 250 nm. The presence of simple molecules can be explained by the the first step of organic matter decomposition of fresh vegetation, further degraded and transformed in humic substances.

References

1. M. Castillo, D. Barceló, *Anal. Chim. Acta*, 426 (2001) 254.
2. J. Labanowski, *Matière Organique Naturelle et Anthropique: Vers une Meilleure Compréhension de sa Réactivité et de sa Caractérisation*. PhD thesis, Université de Limoges, France (2004).
3. M. Crône, *Diagnostic de Sols contaminés par des Hydrocarbures Aromatiques Polycycliques (HAP) à l'Aide de la Spectrophotométrie UV*. PhD thesis, Institut National des Sciences Appliquées de Lyon, France (2000).
4. S.C. Wilson, K.C. Jones, *Environ. Poll.*, 81 (1993) 229.
5. E. Touraud, O. Cloarec, M. Crone, O. Thomas, *Déchets Sci. Tech.*, 25 (2002) 2.
6. AFNOR, Qualité de l'eau, Tome 2, norme NF T 90-114, 1997.
7. P. Prudent, M. Domeizel, C. Massiani, O. Thomas, *Sci. Total Environ.*, 172 (1995) 229.
8. N. Senesi, T.M. Miano, M.R. Provenzano, G. Brunetti, *Soil Sci.*, 152 (1991) 259.
9. H.R. Schulten, Fresenius *J. Anal. Chem.*, 351 (1995) 62.
10. V. Miiki, N. Sensi, K. Hänninen, *Chemosphere*, 34 (1997) 1639.
11. C. Garcia, T. Hernandez, F. Costa, M. Ayuso, *Commun. Sci. Plant Anal.*, 23 (1992) 1501.
12. J.C.G. Esteves Da Silva, A.A.S.C. Marchado, M.A.B.A. Silva, *Wat. Res.*, 32 (1998) 441.
13. S. Deiana, C. Gessa, B. Manunza, R. Rausa, R. Seeber, *Soil Sci.*, 150 (1990) 419.
14. J.L. Morel, B. Nicolradot, *Compost info.*, 22 (1986) 12.

15. L.F. Diaz, G.M. Savage G, L.L Eggerth, C.G. Golueke, *Composting and Recycling of Municipal Waste*, Lewis Publisher (1993).
16. E. Iglesias-Jimenez, V.P. Garcia, *Biol. Wastes*, 27 (1989) 115.
17. M. De Bertoldi, *The Control of the Composting Process and Quality of End Products. In: Composting and Compost Quality Assurance Criteria*. D.V. Jackson, J.M. Merillot, P. l'Hermite. (Eds.), CEC publisher, Luxembourg (1992).
18. M. De Nobili, F. Petrussi, *J. Ferment. Technol.*, 66 (1988) 577.
19. Y. Inbar, Y. Chen, Y. Hadar, H.A.J. Hoitink, *Biocycle*, 12 (1990) 64.
20. K. Sugahara, A. Inoko, *Soil Sci. Plant Nutr.*, 27 (2) (1981) 213.
21. M. Schnitzer, *Soil Biochem.*, 2 (1971) 60.
22. H. De Haan, *Freshwat. Biol.*, 2 (1972) 235.
23. J. Niemeyer, Y. Chen, J.M. Bollag, *Soil Sci. Soc. Am. J.*, 56 (1992) 135.
24. S. Gallot, O. Thomas, *Fresenius J. Anal. Chem.*, 346 (1993) 976.
25. G. Junqua, *Caractérisation Rapide de la Matière Organique de sols et de Sédiments par Spectrophotométrie UV-Visible: Essai de Typologie et Estimation des Paramètres C, N, P*, PhD Thesis, Université de Pau et des Pays de l'Adour, France, (2002).
26. G. Junqua, E. Touraud, O. Thomas, *Déchets Sci. Tech.*, 34 (2004) 14.
27. M. Simonssona, K. Kaiserb, R. Danielssonc, F. Andreuxa, J. Ranger, *Geoderma*, 124 (2005) 157.
28. M. Hassouna, F. Theraulaz, C. Massiani, *Talanta* (2006), Avalaible online: doi:10.1016/j.talanta.2006.05.067.
29. *Standard Methods for the Examination of Water and Wastewater, 10200H Chlorophyll*, 20th edition, A.D. Eaton, L.S. Clesceri, A.E. Greenberg (Eds.), American Public Health Association, Washington, USA, Baltimore, Maryland (1998).

267

CHAPTER 11

UV Spectra Library

S. Spinelli[a], C. Gonzalez[a], O. Thomas[b]

[a]Ecole des Mines d'Alès, 6 avenue de Clavières, 30319 Alès Cedex, France; [b]Université de Sherbrooke, 2500 boulevard de l'université, Sherbrooke, J1K 2R1, Québec, Canada

1. INTRODUCTION

In contrast to the other fields of spectroscopy, there exist very little data related to UV-visible spectrophotometry. One important reference library has been published in 1992 [1], including a lot of spectra of organic compounds with the evolution of molar absorptivity with wavelength. This library is general and not dedicated to water and wastewater applications. One electronic database of 2800 spectra/data sheets (ASCII-format) of some 400 substances subdivided into 19 substance groups is regularly upgraded since 2000 [2]. This electronic database is accessible online and is operated as an open-access database; additional data and information will be added continuously. It aims to gather, from scientific publications, UV-visible spectroscopic data (mainly the evolution of molar absorptivity with wavelength), the traceability of which is not guaranteed. Moreover, this database is more dedicated to atmospheric applications.

The library presented in this chapter includes the spectra of 115 organic and mineral compounds of 11 substance groups. All spectra are acquired following the traceability procedure explained hereafter. In contrast to the previous libraries, the spectra are presented as the variation of absorbance with wavelength, and not the molar absorptivity. A table groups all useful information, including the concentration of the sample solution. This choice of presentation allows users to rapidly show the applicability of UV-visible spectrophotometry for their own applications.

2. SPECTRA ACQUISITION

Spectra acquisition is carried out through different steps:

- Sample selection

  - Compounds (name, n°CAS, M, solubility)
  - Source and purity (supplier, purity)
  - Other parameters (concentration, pH, temperature, solvent)

- Sample preparation

  - Standard solution prepared from two replicates (three if possible)
  - Weighed mass of substance: for each about 500 mg if water soluble
  - Balance (accuracy 0.1 mg)
  - Glassware (cleaning and drying using best available practices)

- ■ Checking pipette accuracy before transfer (if any)
- ■ Checking water quality for solution with acquisition of spectra against air
- ■ Checking solvent quality with acquisition of spectra against air and water (Abs < 0.1 at 200 nm if not another wavelength window)
- ■ Checking buffer quality with acquisition of spectra against air and water (Abs < 0.1 at 200 nm if not another wavelength window)
- ■ Checking if pH conditions are changed by addition of sulphuric acid or sodium hydroxide suppression of the beginning of spectra (for phenolic compounds)
- ■ Temperature (20°C +/− 1°C in a regulated room with temperature registration, no equilibrium time required)

- • Spectrum measurement

- ■ Conditions (spectrophotometer single beam, Anthélie Secomam)
- ■ Spectral bandwidth (fixed 2 nm)
- ■ Wavelength range (200–900 nm)
- ■ Scan speed (1800 nm/min)
- ■ Optical pathlength (10 mm)
- ■ Cell quality (quartz Suprasil® QS)
- ■ Lamp change (350 nm)
- ■ Data pitch (0.5 nm)
- ■ Temperature (20°C +/− 1°C in a regulated room with temperature registration, no equilibrium time required)

More precisely, the general traceability procedure for spectra production is presented in Table 1.

TABLE 1. *Traceability procedure for spectra production*

Step	Nature	Tests	Action
1	**Spectrophotometer check by manufacturer**		
2	**Pharmacopeia check**	**Full procedure**	If N back 1
3	**Check with standards (Hellma)**	WL +/− 0.5 nm Abs +/− 0.01	If N back 2
4	**Spectra acquisition**		
4a	On air (without cell)		Spectrum storage
4b*	On solvent (water or buffer in proportion or organic solvent)		Spectrum storage
4c**	On sample (two replicates)		Spectrum storage
4d	On solvent		Spectrum storage
4e	Test results spectra 4b/4d	diff Abs < 0.005A?	If N back 3
4f	Validation on spectra 4c normalised***	diff Abs < 0.005A?	If N no reporting
5	**Reporting**		
5a	Storage of spectra as ASCII files		Air and solvents Samples normalised
5b	Plotting spectra (GraphPad Prism 2.01) including molecular structure and characteristics	WL 200–450 nm Abs 0–2.5A	Sample (1 sp. or 3 if pH effect)

*Can be repeated in case of pH effect on solutions.
**Can be repeated if no solvent change (maximum five samples).
***Corrected for the mean of solvent contribution and normalised to an exact concentration based on the exact sample weights.
N: test fails.

3. SPECTRA OF COMPOUNDS

3.1. Acids and salts

3.1.1. Acetic acid

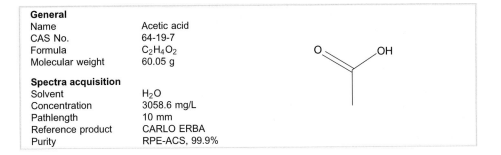

General
Name   Acetic acid
CAS No.   64-19-7
Formula $C_2H_4O_2$
Molecular weight 60.05 g

Spectra acquisition
Solvent H_2O
Concentration   3058.6 mg/L
Pathlength   10 mm
Reference product CARLO ERBA
Purity   RPE-ACS, 99.9%

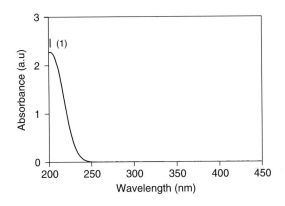

Peak n°	Wavelength (nm)	Absorbance (a.u.)
1	202.0	2.277

3.1.2. Butyric acid

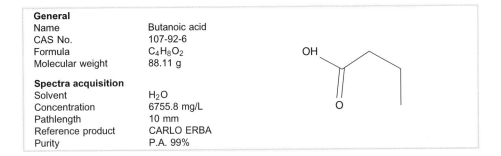

General
Name   Butanoic acid
CAS No.   107-92-6
Formula $C_4H_8O_2$
Molecular weight 88.11 g

Spectra acquisition
Solvent H_2O
Concentration   6755.8 mg/L
Pathlength   10 mm
Reference product CARLO ERBA
Purity   P.A. 99%

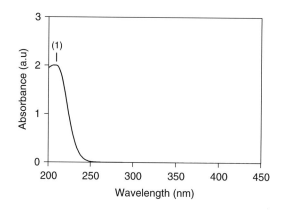

Peak n°	Wavelength (nm)	Absorbance (a.u.)
1	206.3	2.003

3.1.3. EDTA

General	
Name	Ethylenediaminetetraacetic acid disodium salt, dihydrate
CAS No.	6381-92-6
Formula	$C_{10}H_{18}N_2Na_2O_{10}$
Molecular weight	372.24 g
Spectra acquisition	
Solvent	H_2O
Concentration	26.6 mg/L
pH	5.3
Pathlength	10 mm
Reference product	CARLO ERBA
Purity	RPE-ACS 100%

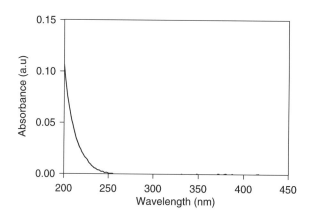

Peak n°	Wavelength (nm)	Absorbance (a.u.)
—	—	—

3.1.4. Formic acid

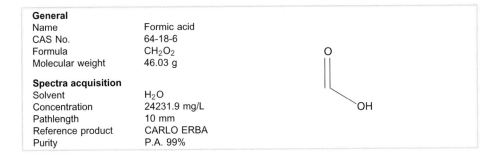

General
Name   Formic acid
CAS No.   64-18-6
Formula CH_2O_2
Molecular weight 46.03 g

Spectra acquisition
Solvent H_2O
Concentration   24231.9 mg/L
Pathlength   10 mm
Reference product CARLO ERBA
Purity   P.A. 99%

Peak n°	Wavelength (nm)	Absorbance (a.u.)
1	225.2	2.490

3.1.5. Oxalic acid

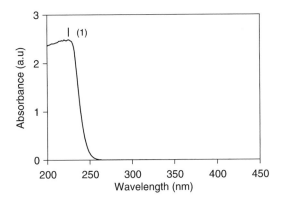

General
Name   Oxalic acid dihydrate
CAS No.   877-24-7
Formula $C_2H_6O_6$
Molecular weight 126.07 g

Spectra acquisition
Solvent H_2O
Concentration   62.7 mg/L
pH   3.3
Pathlength   10 mm
Reference product RDH
Purity   Normadose

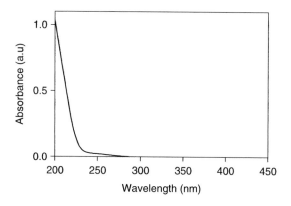

Peak n°	Wavelength (nm)	Absorbance (a.u.)
—	—	—

3.1.6. Propionic acid

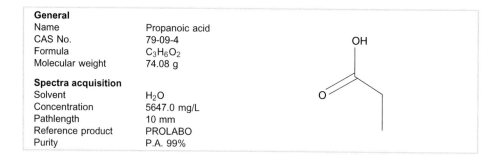

General
Name   Propanoic acid
CAS No.   79-09-4
Formula $C_3H_6O_2$
Molecular weight   74.08 g

Spectra acquisition
Solvent H_2O
Concentration   5647.0 mg/L
Pathlength   10 mm
Reference product   PROLABO
Purity   P.A. 99%

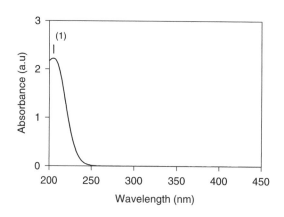

Peak n°	Wavelength (nm)	Absorbance (a.u.)
1	205.0	2.221

3.1.7. Sodium salicylate

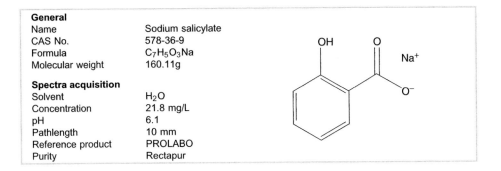

General

Name	Sodium salicylate
CAS No.	578-36-9
Formula	$C_7H_5O_3Na$
Molecular weight	160.11g

Spectra acquisition

Solvent	H_2O
Concentration	21.8 mg/L
pH	6.1
Pathlength	10 mm
Reference product	PROLABO
Purity	Rectapur

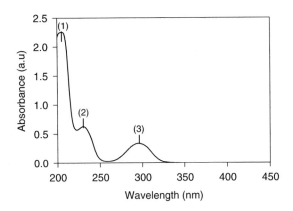

Peak n°	Wavelength (nm)	Absorbance (a.u.)
1	203.7	2.282
2	231.2	0.630
3	298.7	0.336

3.1.8. Potassium sodium tartrate

General
Name   Potassium sodium tartrate tetrahydrate
CAS No.   6381-59-5
Formula $C_4H_{12}NaO_{10}K$
Molecular weight   282.22 g

Spectra acquisition
Solvent H_2O
Concentration   20.2 mg/L
pH   5.9
Pathlength   10 mm
Reference product   ACROS
Purity   P.A.

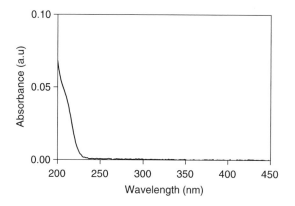

Peak n°	Wavelength (nm)	Absorbance (a.u.)
—	—	—

3.1.9. Valeric acid

General
Name   Pentanoic acid
CAS No.   109-52-4
Formula $C_5H_{10}O_2$
Molecular weight   102.13 g

Spectra acquisition
Solvent H_2O
Concentration   7460 mg/L
Pathlength   10 mm
Reference product   MERCK
Purity   99%

Peak n°	Wavelength (nm)	Absorbance (a.u.)
—	—	—

3.2. Aldehydes and ketones

3.2.1. Acetaldehyde

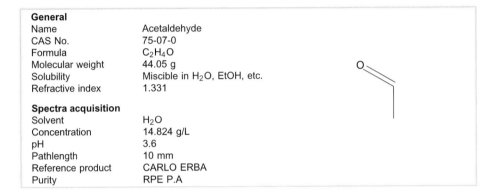

General

Name	Acetaldehyde
CAS No.	75-07-0
Formula	C_2H_4O
Molecular weight	44.05 g
Solubility	Miscible in H_2O, EtOH, etc.
Refractive index	1.331

Spectra acquisition

Solvent	H_2O
Concentration	14.824 g/L
pH	3.6
Pathlength	10 mm
Reference product	CARLO ERBA
Purity	RPE P.A

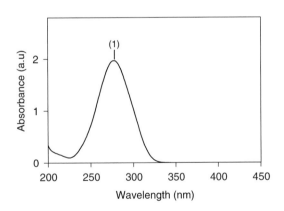

Peak n°	Wavelength (nm)	Absorbance (a.u.)
1	277.1	1.972

3.2.2. Acetone

General
Name	Acetone
CAS No.	67-64-1
Formula	C_3H_6O
Molecular weight	58.08 g
Solubility	Miscible in H_2O, EtOH, etc.
Refractive index	1.359

Spectra acquisition
Solvent	H_2O
Concentration	1.565 g/L
pH	6.2
Pathlength	10 mm
Reference product	CARLO ERBA
Purity	P. spectro (RS)

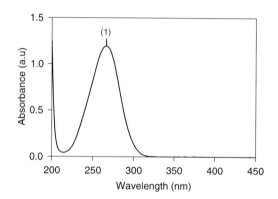

Peak n°	Wavelength (nm)	Absorbance (a.u.)
1	266.1	1.195

Remark: Also registered as solvent; see Section 3.10

3.2.3. Benzaldehyde

General
Name	Benzaldehyde
CAS No.	100-52-7
Formula	C_7H_6O
Molecular weight	106.13 g
Solubility (H_2O)	4 g/L (20°C)
Refractive index	1.546

Spectra acquisition
Solvent	H_2O
Concentration	3.3 mg/L
pH	5.2
Pathlength	10 mm
Reference product	FLUKA
Purity	99% (GC)

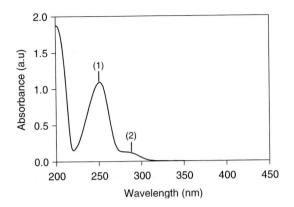

Peak n°	Wavelength (nm)	Absorbance (a.u.)
1	251.2	1.094
2	288.7	0.120

3.2.4. 2-Butanone

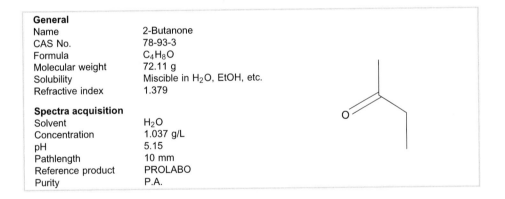

General
Name   2-Butanone
CAS No.   78-93-3
Formula C_4H_8O
Molecular weight   72.11 g
Solubility   Miscible in H_2O, EtOH, etc.
Refractive index   1.379

Spectra acquisition
Solvent H_2O
Concentration   1.037 g/L
pH   5.15
Pathlength   10 mm
Reference product   PROLABO
Purity   P.A.

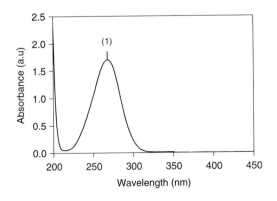

Peak n°	Wavelength (nm)	Absorbance (a.u.)
1	268.1	1.703

3.2.5.   Butyraldehyde

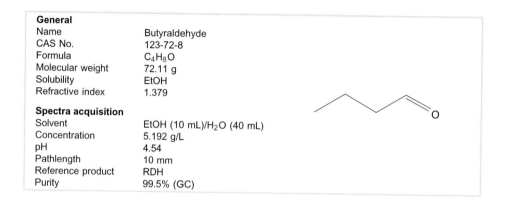

General
Name	Butyraldehyde
CAS No.	123-72-8
Formula	C_4H_8O
Molecular weight	72.11 g
Solubility	EtOH
Refractive index	1.379

Spectra acquisition
Solvent	EtOH (10 mL)/H_2O (40 mL)
Concentration	5.192 g/L
pH	4.54
Pathlength	10 mm
Reference product	RDH
Purity	99.5% (GC)

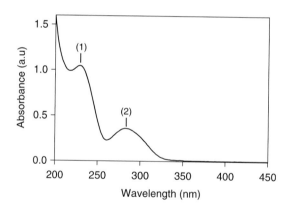

Peak n°	Wavelength (nm)	Absorbance (a.u.)
1	228.9	1.047
2	282.9	0.362

3.2.6. Diisobutylketone

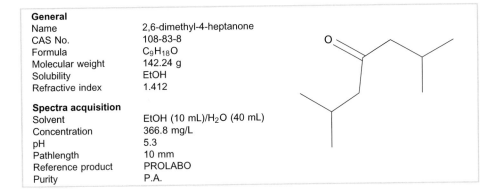

General
Name	2,6-dimethyl-4-heptanone
CAS No.	108-83-8
Formula	$C_9H_{18}O$
Molecular weight	142.24 g
Solubility	EtOH
Refractive index	1.412

Spectra acquisition
Solvent	EtOH (10 mL)/H_2O (40 mL)
Concentration	366.8 mg/L
pH	5.3
Pathlength	10 mm
Reference product	PROLABO
Purity	P.A.

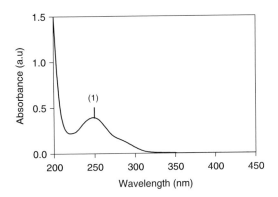

Peak n°	Wavelength (nm)	Absorbance (a.u.)
1	248.2	0.394

3.2.7. Formaldehyde

General
Name	Formaldehyde
CAS No.	50-00-0
Formula	CH_2O
Molecular weight	30.03 g
Solubility	EtOH
Refractive index	1.412

Spectra acquisition
Solvent	Pure reagent
Concentration	400 g/L
pH	—
Pathlength	10 mm
Reference product	CARLO ERBA
Purity	40% + 10% MeOH

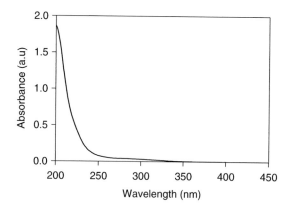

Peak n°	Wavelength (nm)	Absorbance (a.u.)
1	—	—

3.2.8. Isobutyl methyl ketone

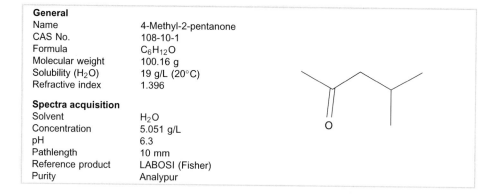

General
Name	4-Methyl-2-pentanone
CAS No.	108-10-1
Formula	$C_6H_{12}O$
Molecular weight	100.16 g
Solubility (H_2O)	19 g/L (20°C)
Refractive index	1.396

Spectra acquisition
Solvent	H_2O
Concentration	5.051 g/L
pH	6.3
Pathlength	10 mm
Reference product	LABOSI (Fisher)
Purity	Analypur

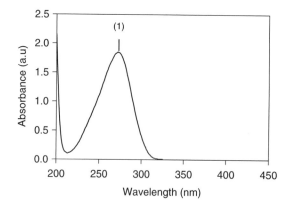

Peak n°	Wavelength (nm)	Absorbance (a.u.)
1	273.2	1.846

3.3. Amines and related compounds

3.3.1. Aniline

General

Name	Aniline
CAS No.	62-53-3
Formula	C_6H_7N
Molecular weight	93.13 g
Solubility (H_2O)	34 g/L (20°C)
Refractive index	1.58

Spectra acquisition

Solvent	H_2O
Concentration	20.1 mg/L
pH	6.3
Pathlength	10 mm
Reference product	ALLTECH
Purity	99%

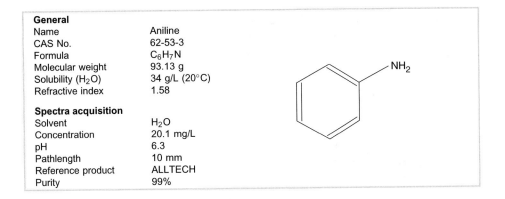

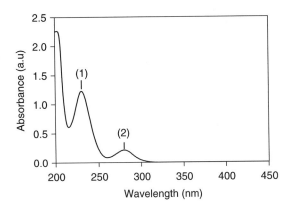

Peak n°	Wavelength (nm)	Absorbance (a.u.)
1	230.3	1.232
2	279.5	0.215

3.3.2. p-Anisidine

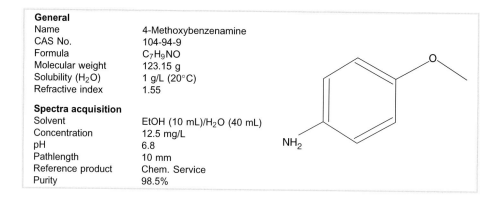

General
Name   4-Methoxybenzenamine
CAS No.   104-94-9
Formula C_7H_9NO
Molecular weight   123.15 g
Solubility (H_2O) 1 g/L (20°C)
Refractive index   1.55

Spectra acquisition
Solvent   EtOH (10 mL)/H_2O (40 mL)
Concentration   12.5 mg/L
pH   6.8
Pathlength   10 mm
Reference product   Chem. Service
Purity   98.5%

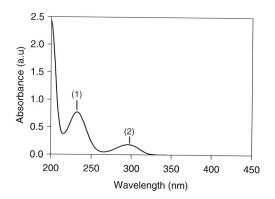

Peak n°	Wavelength (nm)	Absorbance (a.u.)
1	232.3	0.770
2	295.7	0.184

3.3.3. 2-Chloroaniline

General
Name   2-Chloroaniline
CAS No.   95-51-2
Formula C_6H_6ClN
Molecular weight   127.57 g
Solubility   EtOH
Refractive index   1.59

Spectra acquisition
Solvent   EtOH/H_2O
Concentration   10.0 mg/L
pH   5.3
Pathlength   10 mm
Reference product   ACROS
Purity   98%

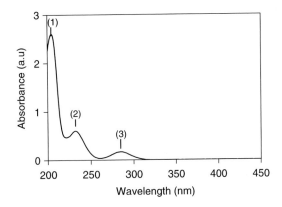

Peak n°	Wavelength (nm)	Absorbance (a.u.)
1	204.7	2.607
2	232.0	0.602
3	284.6	0.169

3.3.4. 4-Chloroaniline

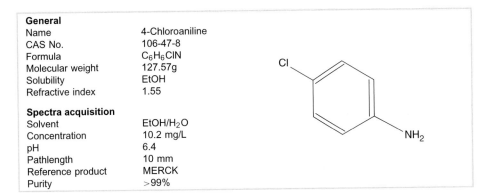

General
Name	4-Chloroaniline
CAS No.	106-47-8
Formula	C_6H_6ClN
Molecular weight	127.57g
Solubility	EtOH
Refractive index	1.55

Spectra acquisition
Solvent	EtOH/H_2O
Concentration	10.2 mg/L
pH	6.4
Pathlength	10 mm
Reference product	MERCK
Purity	>99%

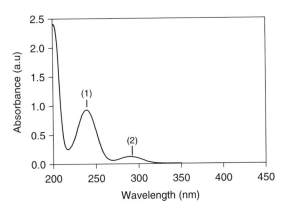

Peak n°	Wavelength (nm)	Absorbance (a.u.)
1	238.9	0.911
2	290.2	0.121

3.3.5. 2-Chloro-4-methylaniline

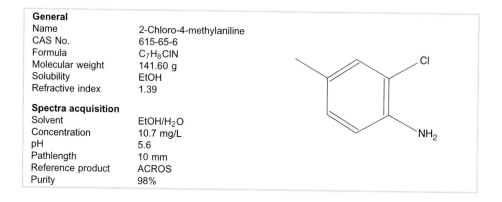

General
Name   2-Chloro-4-methylaniline
CAS No.   615-65-6
Formula C_7H_8ClN
Molecular weight   141.60 g
Solubility   EtOH
Refractive index   1.39

Spectra acquisition
Solvent $EtOH/H_2O$
Concentration   10.7 mg/L
pH   5.6
Pathlength   10 mm
Reference product   ACROS
Purity   98%

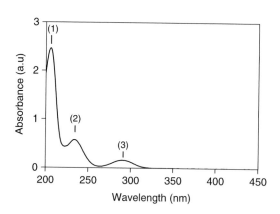

Peak n°	Wavelength (nm)	Absorbance (a.u.)
1	204.7	2.607
2	232.0	0.602
3	284.6	0.169

3.3.6. 3,4-Dichloroaniline

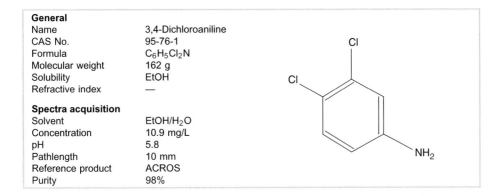

General	
Name	3,4-Dichloroaniline
CAS No.	95-76-1
Formula	$C_6H_5Cl_2N$
Molecular weight	162 g
Solubility	EtOH
Refractive index	—

Spectra acquisition	
Solvent	EtOH/H_2O
Concentration	10.9 mg/L
pH	5.8
Pathlength	10 mm
Reference product	ACROS
Purity	98%

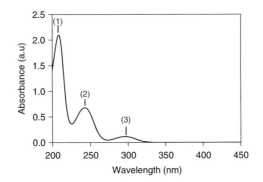

Peak n°	Wavelength (nm)	Absorbance (a.u.)
1	208.5	2.100
2	242.4	0.677
3	295.9	0.115

3.3.7. Diethylamine

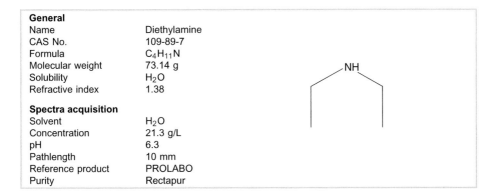

General	
Name	Diethylamine
CAS No.	109-89-7
Formula	$C_4H_{11}N$
Molecular weight	73.14 g
Solubility	H_2O
Refractive index	1.38

Spectra acquisition	
Solvent	H_2O
Concentration	21.3 g/L
pH	6.3
Pathlength	10 mm
Reference product	PROLABO
Purity	Rectapur

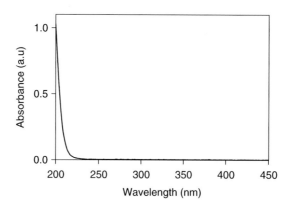

Peak n°	Wavelength (nm)	Absorbance (a.u.)
—	—	—

3.3.8.   Diethanolamine

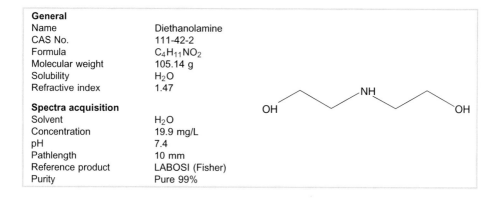

General
Name   Diethanolamine
CAS No.   111-42-2
Formula $C_4H_{11}NO_2$
Molecular weight   105.14 g
Solubility H_2O
Refractive index   1.47

Spectra acquisition
Solvent H_2O
Concentration   19.9 mg/L
pH   7.4
Pathlength   10 mm
Reference product   LABOSI (Fisher)
Purity   Pure 99%

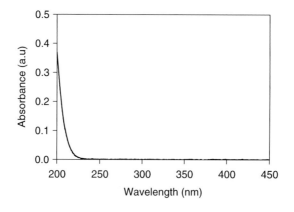

Peak n°	Wavelength (nm)	Absorbance (a.u.)
—	—	—

3.3.9. Glutamic acid

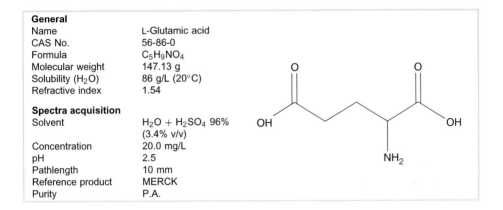

General
Name   L-Glutamic acid
CAS No.   56-86-0
Formula $C_5H_9NO_4$
Molecular weight   147.13 g
Solubility (H_2O)   86 g/L (20°C)
Refractive index   1.54

Spectra acquisition
Solvent $H_2O + H_2SO_4$ 96%
                      (3.4% v/v)
Concentration   20.0 mg/L
pH   2.5
Pathlength   10 mm
Reference product   MERCK
Purity   P.A.

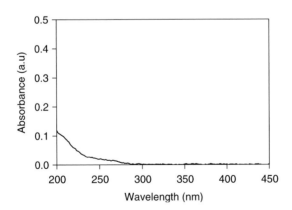

Peak n°	Wavelength (nm)	Absorbance (a.u.)
—	—	—

3.3.10. Glycine

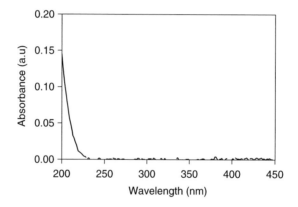

General

Name	Aminoacetic acid
N° CAS of glycine	56-40-6
Formula	$C_2H_5NO_2$
Molecular weight	75.07 g
Solubility (H_2O)	250 g/L (20°C)
Refractive index	—

Spectra acquisition

Solvent	H_2O
Concentration	20.0 g/L
pH	6.85
Pathlength	10 mm
Reference product	RDH
Purity	P.A.

Peak n°	Wavelength (nm)	Absorbance (a.u.)
—	—	—

3.3.11. 4-Nitroaniline

General

Name	4-Nitroaniline
CAS No.	100-01-61
Formula	$C_6H_6N_2O_2$
Molecular weight	138.13 g
Solubility (H_2O)	0.8 g/L (20°C)
Refractive index	—

Spectra acquisition

Solvent	EtOH/H_2O
Concentration	10.0 mg/L
pH	5.6
Pathlength	10 mm
Reference product	RDH
Purity	99.5%

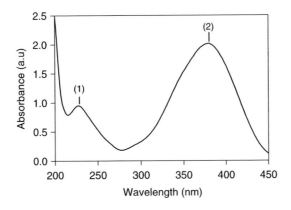

Peak n°	Wavelength (nm)	Absorbance (a.u.)
1	227.6	0.952
2	379.7	2.017

3.3.12. *m-Toluidine*

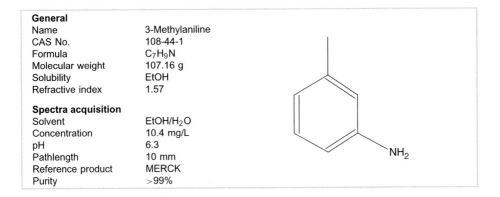

General

Name	3-Methylaniline
CAS No.	108-44-1
Formula	C_7H_9N
Molecular weight	107.16 g
Solubility	EtOH
Refractive index	1.57

Spectra acquisition

Solvent	EtOH/H_2O
Concentration	10.4 mg/L
pH	6.3
Pathlength	10 mm
Reference product	MERCK
Purity	>99%

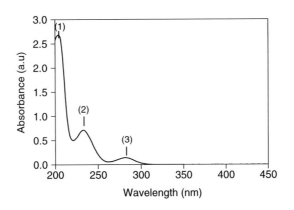

Peak n°	Wavelength (nm)	Absorbance (a.u.)
1	203.9	2.688
2	232.9	0.728
3	281.4	0.140

3.3.13. p-Toluidine

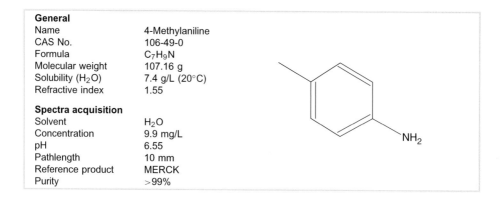

General

Name	4-Methylaniline
CAS No.	106-49-0
Formula	C_7H_9N
Molecular weight	107.16 g
Solubility (H_2O)	7.4 g/L (20°C)
Refractive index	1.55

Spectra acquisition

Solvent	H_2O
Concentration	9.9 mg/L
pH	6.55
Pathlength	10 mm
Reference product	MERCK
Purity	>99%

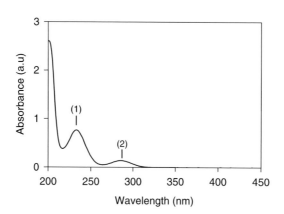

Peak n°	Wavelength (nm)	Absorbance (a.u.)
1	232.9	0.765
2	285.2	0.139

3.3.14. Tyrosine

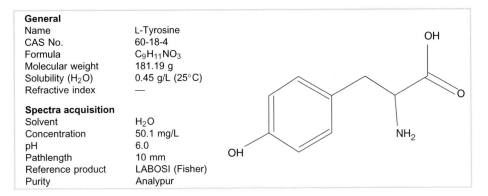

General
Name	L-Tyrosine
CAS No.	60-18-4
Formula	$C_9H_{11}NO_3$
Molecular weight	181.19 g
Solubility (H_2O)	0.45 g/L (25°C)
Refractive index	—

Spectra acquisition
Solvent	H_2O
Concentration	50.1 mg/L
pH	6.0
Pathlength	10 mm
Reference product	LABOSI (Fisher)
Purity	Analypur

Peak n°	Wavelength (nm)	Absorbance (a.u.)
1	224.3	1.686
2	274.8	0.300

3.4. Benzene and related compounds

3.4.1. Benzene

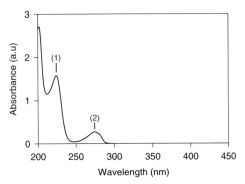

General
Name	Benzene
CAS No.	71-43-2
Formula	C_6H_6
Molecular weight	78.11 g
Solubility (H_2O)	700 mg/L (20°C)
Refractive index	1.50

Spectra acquisition
Solvent	CH_3CN/H_2O (5% v/v)
Concentration	587.3 mg/L
pH	5.8
Pathlength	10 mm
Reference product	RDH
Purity	P.A.

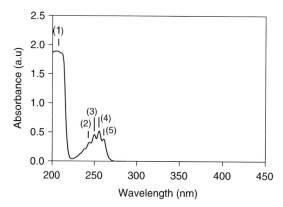

Peak n°	Wavelength (nm)	Absorbance (a.u.)
1	204.9	1.891
2	244.1	0.323
3	249.9	0.457
4	255.6	0.527
5	260.9	0.383

3.4.2. Chlorobenzene

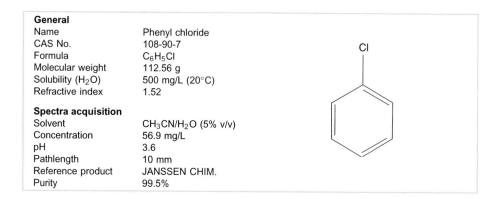

General
Name   Phenyl chloride
CAS No.   108-90-7
Formula C_6H_5Cl
Molecular weight   112.56 g
Solubility (H_2O)   500 mg/L (20°C)
Refractive index   1.52

Spectra acquisition
Solvent CH_3CN/H_2O (5% v/v)
Concentration   56.9 mg/L
pH   3.6
Pathlength   10 mm
Reference product   JANSSEN CHIM.
Purity   99.5%

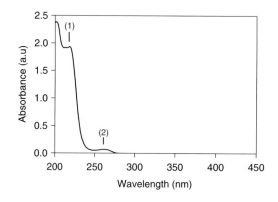

Peak n°	Wavelength (nm)	Absorbance (a.u.)
1	218.4	2.384
2	261.0	0.071

3.4.3. Ethylbenzene

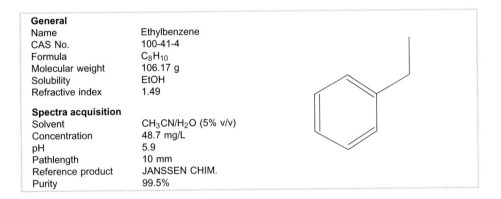

General
Name   Ethylbenzene
CAS No.   100-41-4
Formula C_8H_{10}
Molecular weight   106.17 g
Solubility   EtOH
Refractive index   1.49

Spectra acquisition
Solvent CH_3CN/H_2O (5% v/v)
Concentration   48.7 mg/L
pH   5.9
Pathlength   10 mm
Reference product   JANSSEN CHIM.
Purity   99.5%

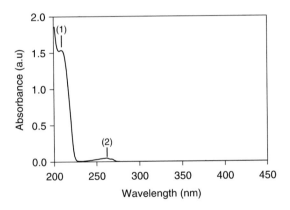

Peak n°	Wavelength (nm)	Absorbance (a.u.)
1	209.0	1.540
2	261.5	0.050

3.4.4. Toluene

General
Name   Methyl benzene
CAS No.   108-88-3
Formula C_7H_8
Molecular weight   92.14 g
Solubility   EtOH
Refractive index   1.49

Spectra acquisition
Solvent CH_3CN/H_2O (5% v/v)
Concentration   41.5 mg/L
pH   5.7
Pathlength   10 mm
Reference product   CARLO ERBA
Purity   99.5%

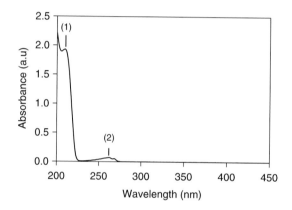

Peak n°	Wavelength (nm)	Absorbance (a.u.)
1	210.4	1.950
2	268.1	0.062

3.4.5. m-Xylene

General
Name   1,3-Dimethyl benzene
CAS No.   108-38-3
Formula C_8H_{10}
Molecular weight   106.17 g
Solubility (H_2O)   200 mg/L (20°C)
Refractive index   1.49

Spectra acquisition
Solvent $MeOH/H_2O$ (0.5% v/v)
Concentration   42.3 mg/L
pH   5.9
Pathlength   10 mm
Reference product   ACROS ORGANICS
Purity   Spectrograde

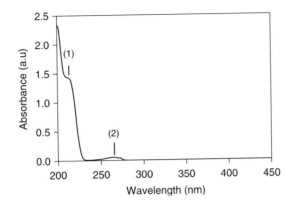

Peak n°	Wavelength (nm)	Absorbance (a.u.)
1	212.9	1.437
2	271.4	0.057

3.4.6.   o-Xylene

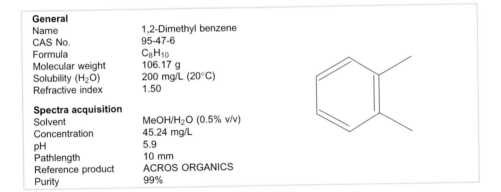

General
Name	1,2-Dimethyl benzene
CAS No.	95-47-6
Formula	C_8H_{10}
Molecular weight	106.17 g
Solubility (H_2O)	200 mg/L (20°C)
Refractive index	1.50

Spectra acquisition
Solvent	MeOH/H_2O (0.5% v/v)
Concentration	45.24 mg/L
pH	5.9
Pathlength	10 mm
Reference product	ACROS ORGANICS
Purity	99%

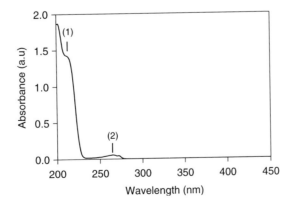

Peak n°	Wavelength (nm)	Absorbance (a.u.)
1	211.3	1.427
2	264.9	0.058

3.4.7. p-Xylene

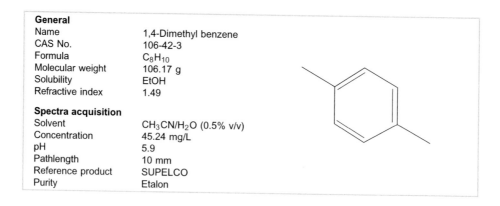

General

Name	1,4-Dimethyl benzene
CAS No.	106-42-3
Formula	C_8H_{10}
Molecular weight	106.17 g
Solubility	EtOH
Refractive index	1.49

Spectra acquisition

Solvent	CH_3CN/H_2O (0.5% v/v)
Concentration	45.24 mg/L
pH	5.9
Pathlength	10 mm
Reference product	SUPELCO
Purity	Etalon

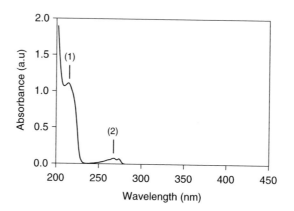

Peak n°	Wavelength (nm)	Absorbance (a.u.)
1	214.6	1.110
2	267.8	0.079

3.5. Phenol and related compounds

3.5.1. Phenol

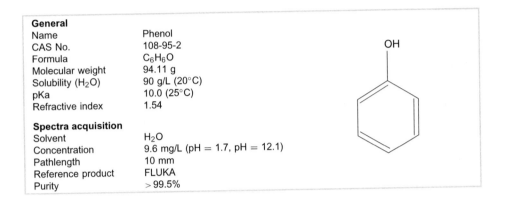

General
Name	Phenol
CAS No.	108-95-2
Formula	C_6H_6O
Molecular weight	94.11 g
Solubility (H_2O)	90 g/L (20°C)
pKa	10.0 (25°C)
Refractive index	1.54

Spectra acquisition
Solvent	H_2O
Concentration	9.6 mg/L (pH = 1.7, pH = 12.1)
Pathlength	10 mm
Reference product	FLUKA
Purity	> 99.5%

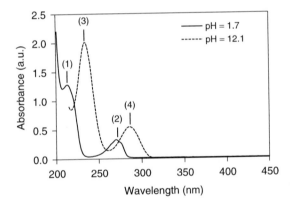

Peak n°	Wavelength (nm)	Absorbance (a.u.)
1	213.0	1.285
2	270.3	0.331
3	234.5	2.014
4	285.6	0.550

3.5.2. 4-Chloro-3-methylphenol

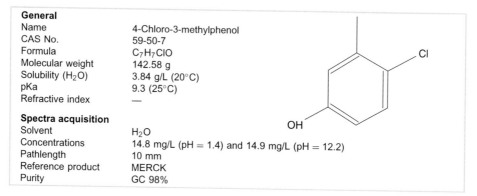

General	
Name	4-Chloro-3-methylphenol
CAS No.	59-50-7
Formula	C_7H_7ClO
Molecular weight	142.58 g
Solubility (H_2O)	3.84 g/L (20°C)
pKa	9.3 (25°C)
Refractive index	—

Spectra acquisition	
Solvent	H_2O
Concentrations	14.8 mg/L (pH = 1.4) and 14.9 mg/L (pH = 12.2)
Pathlength	10 mm
Reference product	MERCK
Purity	GC 98%

Peak n°	Wavelength (nm)	Absorbance (a.u.)
1	226.7	0.745
2	279.4	0.163
3	243.7	1.121
4	297.2	0.270

3.5.3. 2-Chlorophenol

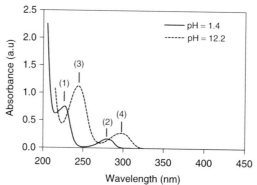

General	
Name	2-Chlorophenol
CAS No.	95-57-8
Formula	C_6H_5ClO
Molecular weight	128.57 g
Solubility	EtOH
pKa	8.6 (25°C)
Refractive index	1.55

Spectra acquisition	
Solvent	EtOH/H_2O (0.2% v/v)
Concentrations	19.2 mg/L (pH = 1.4) and 21.4 mg/L (pH = 12.1)
Pathlength	10 mm
Reference product	RDH
Purity	GC 98%

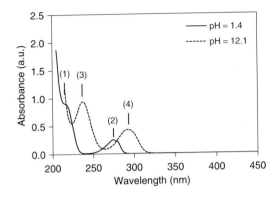

Peak n°	Wavelength (nm)	Absorbance (a.u.)
1	213.9	0.918
2	274.0	0.208
3	237.1	0.942
4	292.4	0.434

3.5.4. 3-Chlorophenol

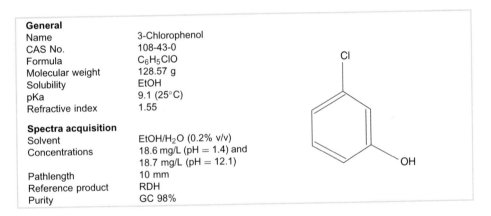

General

Name	3-Chlorophenol
CAS No.	108-43-0
Formula	C_6H_5ClO
Molecular weight	128.57 g
Solubility	EtOH
pKa	9.1 (25°C)
Refractive index	1.55

Spectra acquisition

Solvent	EtOH/H_2O (0.2% v/v)
Concentrations	18.6 mg/L (pH = 1.4) and 18.7 mg/L (pH = 12.1)
Pathlength	10 mm
Reference product	RDH
Purity	GC 98%

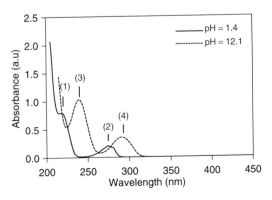

Peak n°	Wavelength (nm)	Absorbance (a.u.)
1	218.1	0.795
2	274.1	0.204
3	239.8	1.033
4	290.4	0.357

3.5.5. 4-Chlorophenol

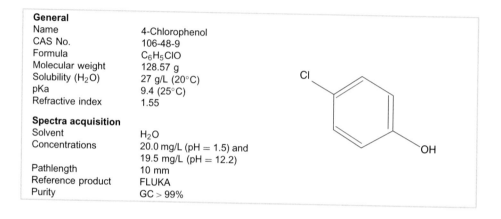

General

Name	4-Chlorophenol
CAS No.	106-48-9
Formula	C_6H_5ClO
Molecular weight	128.57 g
Solubility (H_2O)	27 g/L (20°C)
pKa	9.4 (25°C)
Refractive index	1.55

Spectra acquisition

Solvent	H_2O
Concentrations	20.0 mg/L (pH = 1.5) and 19.5 mg/L (pH = 12.2)
Pathlength	10 mm
Reference product	FLUKA
Purity	GC > 99%

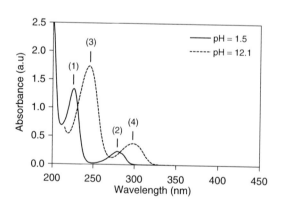

Peak n°	Wavelength (nm)	Absorbance (a.u.)
1	226.0	1.337
2	279.3	0.236
3	244.8	1.739
4	298.0	0.378

3.5.6. m-Cresol

General

Name	3-Methylphenol
CAS No.	108-39-4
Formula	C_7H_8O
Molecular weight	108.14 g
Solubility (H_2O)	25 g/L (20°C)
pKa	10.1 (25°C)
Refractive index	1.54

Spectra acquisition

Solvent	H_2O
Concentrations	20.1 mg/L (pH = 1.4) and 20.0 mg/L (pH = 12.1)
Pathlength	10 mm
Reference product	MERCK
Purity	GC 98%

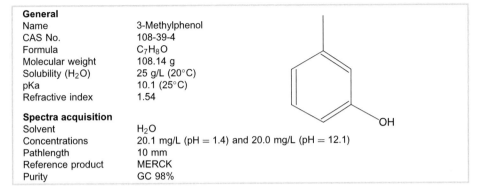

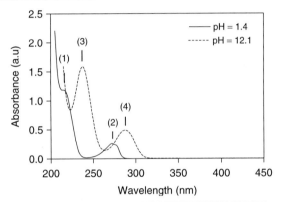

Peak n°	Wavelength (nm)	Absorbance (a.u.)
1	214.8	1.189
2	272.1	0.280
3	237.5	1.568
4	287.5	0.498

3.5.7. o-Cresol

General

Name	2-Methylphenol
CAS No.	95-48-7
Formula	C_7H_8O
Molecular weight	108.14 g
Solubility (H_2O)	25 g/L (20°C)
pKa	10.3 (25°C)
Refractive index	1.54

Spectra acquisition

Solvent	H_2O
Concentrations	19.7 mg/L (pH = 1.4) and 19.9 mg/L (pH = 12.1)
Pathlength	10 mm
Reference product	MERCK
Purity	GC 99%

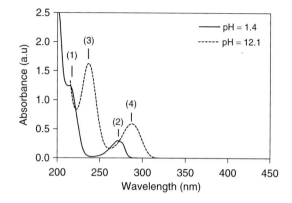

Peak n°	Wavelength (nm)	Absorbance (a.u.)
1	216.2	1.260
2	271.3	0.298
3	236.6	1.622
4	287.4	0.591

3.5.8. p-Cresol

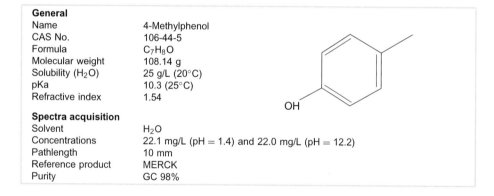

General

Name	4-Methylphenol
CAS No.	106-44-5
Formula	C_7H_8O
Molecular weight	108.14 g
Solubility (H_2O)	25 g/L (20°C)
pKa	10.3 (25°C)
Refractive index	1.54

Spectra acquisition

Solvent	H_2O
Concentrations	22.1 mg/L (pH = 1.4) and 22.0 mg/L (pH = 12.2)
Pathlength	10 mm
Reference product	MERCK
Purity	GC 98%

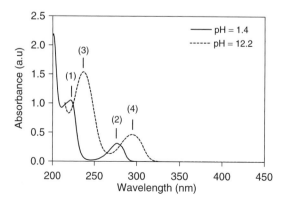

Peak n°	Wavelength (nm)	Absorbance (a.u.)
1	221.9	1.053
2	276.9	0.312
3	236.7	1.539
4	294.4	0.465

3.5.9. 4,5-Dichlorocatechol

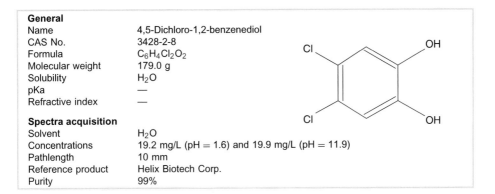

General
Name   4,5-Dichloro-1,2-benzenediol
CAS No.   3428-2-8
Formula $C_6H_4Cl_2O_2$
Molecular weight   179.0 g
Solubility H_2O
pKa   —
Refractive index   —

Spectra acquisition
Solvent H_2O
Concentrations   19.2 mg/L (pH = 1.6) and 19.9 mg/L (pH = 11.9)
Pathlength   10 mm
Reference product   Helix Biotech Corp.
Purity   99%

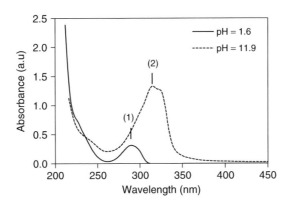

Peak n°	Wavelength (nm)	Absorbance (a.u.)
1	289.9	0.314
2	315.1	1.333

3.5.10.   2,3-Dichlorophenol

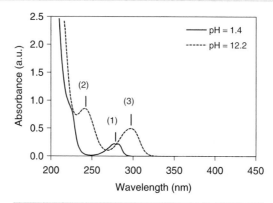

General
Name	2,3-Dichlorophenol
CAS No.	576-24-9
Formula	$C_6H_4Cl_2O$
Molecular weight	163.0 g
Solubility	EtOH
pKa	7.4 (25°C)
Refractive index	1.54

Spectra acquisition
Solvent	EtOH/H_2O (2% v/v)
Concentrations	20.3 mg/L (pH = 1.4) and 21.0 mg/L (pH = 12.2)
Pathlength	10 mm
Reference product	MERCK
Purity	GC 98%

Peak n°	Wavelength (nm)	Absorbance (a.u.)
1	276.7	0.223
2	241.1	0.855
3	297.1	0.494

3.5.11.   2,4-Dichlorophenol

General
Name	2,4-Dichlorophenol
CAS No.	120-83-2
Formula	$C_6H_4Cl_2O$
Molecular weight	163.0 g
Solubility (H_2O)	25 g/L (20°C)
pKa	—
Refractive index	—

Spectra acquisition
Solvent	H_2O
Concentrations	7.8 mg/L (pH = 1.6) and 7.9 mg/L (pH = 12.0)
Pathlength	10 mm
Reference product	RDH
Purity	Pestanal

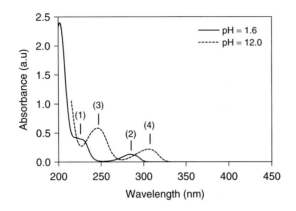

Peak n°	Wavelength (nm)	Absorbance (a.u.)
1	221.2	0.478
2	283.2	0.133
3	245.9	0.587
4	306.1	0.220

3.5.12. 2,5-Dimethylphenol

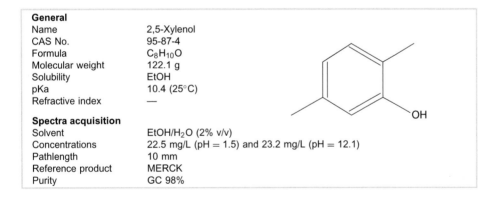

General

Name	2,5-Xylenol
CAS No.	95-87-4
Formula	$C_8H_{10}O$
Molecular weight	122.1 g
Solubility	EtOH
pKa	10.4 (25°C)
Refractive index	—

Spectra acquisition

Solvent	EtOH/H$_2$O (2% v/v)
Concentrations	22.5 mg/L (pH = 1.5) and 23.2 mg/L (pH = 12.1)
Pathlength	10 mm
Reference product	MERCK
Purity	GC 98%

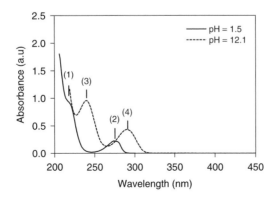

Peak n°	Wavelength (nm)	Absorbance (a.u.)
1	218.0	0.932
2	274.3	0.225
3	239.7	0.959
4	290.3	0.433

3.5.13. 4,6-Dinitro-2-methylphenol

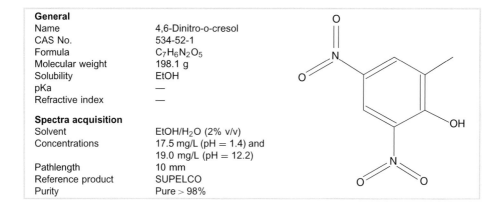

General
Name	4,6-Dinitro-o-cresol
CAS No.	534-52-1
Formula	$C_7H_6N_2O_5$
Molecular weight	198.1 g
Solubility	EtOH
pKa	—
Refractive index	—

Spectra acquisition
Solvent	EtOH/H_2O (2% v/v)
Concentrations	17.5 mg/L (pH = 1.4) and 19.0 mg/L (pH = 12.2)
Pathlength	10 mm
Reference product	SUPELCO
Purity	Pure > 98%

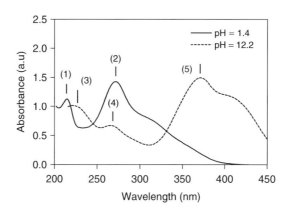

Peak n°	Wavelength (nm)	Absorbance (a.u.)
1	214.8	1.133
2	271.5	1.429
3	220.9	1.015
4	265.7	0.671
5	371.1	1.495

3.5.14. 2-Nitrophenol

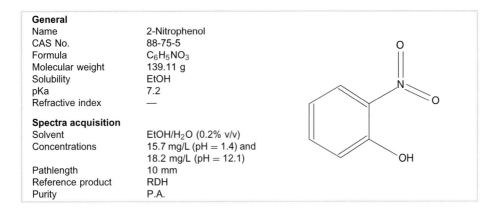

General

Name	2-Nitrophenol
CAS No.	88-75-5
Formula	$C_6H_5NO_3$
Molecular weight	139.11 g
Solubility	EtOH
pKa	7.2
Refractive index	—

Spectra acquisition

Solvent	EtOH/H_2O (0.2% v/v)
Concentrations	15.7 mg/L (pH = 1.4) and 18.2 mg/L (pH = 12.1)
Pathlength	10 mm
Reference product	RDH
Purity	P.A.

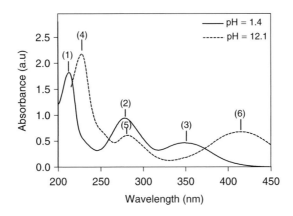

Peak n°	Wavelength (nm)	Absorbance (a.u.)
1	212.6	1.822
2	278.1	0.943
3	348.5	0.465
4	227.6	2.182
5	281.3	0.609
6	414.0	0.674

3.5.15. 3-Nitrophenol

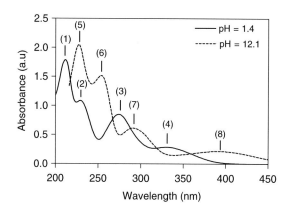

General	
Name	3-Nitrophenol
CAS No.	554-84-7
Formula	$C_6H_5NO_3$
Molecular weight	139.11 g
Solubility (H_2O)	30 g/L (40°C)
pKa	8.3
Refractive index	—
Spectra acquisition	
Solvent	EtOH/H_2O (0.2% v/v)
Concentrations	17.2 mg/L (pH = 1.4) and
	20.7 mg/L (pH = 12.1)
Pathlength	10 mm
Reference product	RDH
Purity	Indicator

Peak n°	Wavelength (nm)	Absorbance (a.u.)
1	211.6	1.800
2	228.8	1.094
3	273.9	0.857
4	320.2	0.289
5	227.0	2.051
6	253.4	1.516
7	290.3	0.615
8	387.4	0.218

3.5.16. 4-Nitrophenol

General

Name	4-Nitrophenol
CAS No.	100-02-7
Formula	$C_6H_5NO_3$
Molecular weight	139.11 g
Solubility	H_2O
pKa	7.15
Refractive index	—

Spectra acquisition

Solvent	EtOH/H_2O (0.25% v/v)
Concentrations	22.2 mg/L (pH = 1.7) and 23.1 mg/L (pH = 12.1)
Pathlength	10 mm
Reference product	SUPELCO
Purity	P.A.

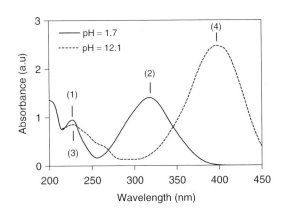

Peak n°	Wavelength (nm)	Absorbance (a.u.)
1	226.2	0.966
2	317.6	1.094
3	227.1	0.854
4	397.0	2.475

3.5.17. Pentachlorophenol

General

Name	Pentachlorophenol
CAS No.	87-86-5
Formula	C_6HCl_5O
Molecular weight	266.34 g
Solubility (H_2O)	20 mg/L (20°C)
pKa	—
Refractive index	—

Spectra acquisition

Solvent	EtOH/H_2O (0.25% v/v)
Concentrations	15.0 mg/L (pH = 1.4) and
	14.5 mg/L (pH = 12.0)
Pathlength	10 mm
Reference product	CARLO ERBA
Purity	RPE (99%)

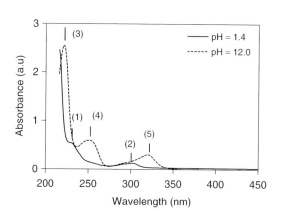

Peak n°	Wavelength (nm)	Absorbance (a.u.)
1	230.1	0.539
2	301.8	0.126
3	221.4	2.555
4	250.5	0.593
5	319.8	0.302

3.5.18. Pyrocatechol

General

Name	1,2-Benzenediol
CAS No.	120-80-9
Formula	$C_6H_6O_2$
Molecular weight	110.11 g
Solubility (H_2O)	435 mg/L (20°C)
pKa	9.3
Refractive index	1.60

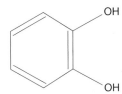

Spectra acquisition

Solvent	EtOH/H_2O (0.2% v/v)
Concentrations	17.6 mg/L (pH = 1.4) and 19.0 mg/L (pH = 12.0)
Pathlength	10 mm
Reference product	CARLO ERBA
Purity	RPE (99%)

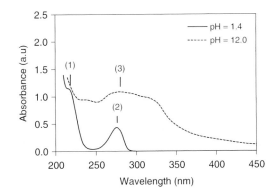

Peak n°	Wavelength (nm)	Absorbance (a.u.)
1	216.2	1.148
2	275.4	0.433
3	280.0	1.087

3.5.19. 2-Tert-butyl-4-methylphenol

General

Name	2-Tert-butyl-4-methylphenol
CAS No.	2409-55-4
Formula	$C_{11}H_{16}O$
Molecular weight	164.25 g
Solubility	EtOH
pKa	—
Refractive index	1.49

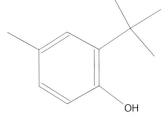

Spectra acquisition

Solvent	EtOH/H_2O (0.2% v/v)
Concentrations	17.6 mg/L (pH = 1.6) and 21.0 mg/L (pH = 12.0)
Pathlength	10 mm
Reference product	ALDRICH
Purity	99%

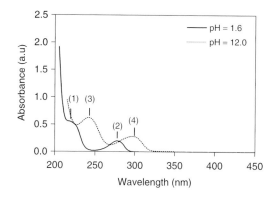

Peak n°	Wavelength (nm)	Absorbance (a.u.)
1	217.1	0.559
2	277.6	0.197
3	241.9	0.624
4	296.6	0.280

3.5.20. 2,4,6-Trichlorophenol

General

Name	2,4,6-Trichlorophenol
CAS No.	88-06-2
Formula	$C_6H_3Cl_3O$
Molecular weight	197.45 g
Solubility	EtOH
pKa	6.0
Refractive index	1.49

Spectra acquisition

Solvent	EtOH/H_2O (0.25% v/v)
Concentrations	18.5 mg/L (pH = 1.4) and
	21.8 mg/L (pH = 12.2)
Pathlength	10 mm
Reference product	SUPELCO
Purity	Pure

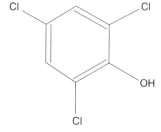

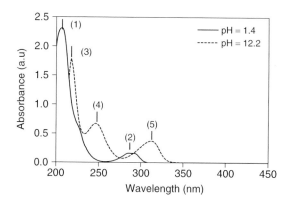

Peak n°	Wavelength (nm)	Absorbance (a.u.)
1	205.4	2.310
2	285.9	0.164
3	217.9	1.782
4	246.3	0.670
5	312.0	0.371

3.5.21. 2,4,6-Trimethylphenol

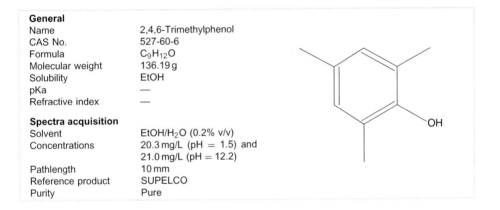

General
Name	2,4,6-Trimethylphenol
CAS No.	527-60-6
Formula	$C_9H_{12}O$
Molecular weight	136.19 g
Solubility	EtOH
pKa	—
Refractive index	—

Spectra acquisition
Solvent	EtOH/H_2O (0.2% v/v)
Concentrations	20.3 mg/L (pH = 1.5) and 21.0 mg/L (pH = 12.2)
Pathlength	10 mm
Reference product	SUPELCO
Purity	Pure

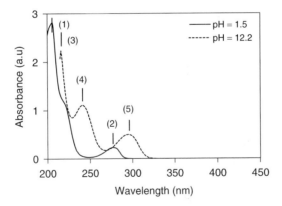

Peak n°	Wavelength (nm)	Absorbance (a.u.)
1	205.6	2.819
2	277.1	0.235
3	216.1	2.238
4	241.1	1.108
5	295.2	0.504

3.6. Phthalates

3.6.1. Butyl benzyl phthalate

General

Name	Butyl benzyl phthalate
CAS No.	85-68-7
Formula	$C_{19}H_{20}O_4$
Molecular weight	312.36 g
Solubility (H_2O)	2.7 mg/L (20°C)
Refractive index	—

Spectra acquisition

Solvent	MetOH/H_2O (20% v/v)
Concentrations	27.1 mg/L
Pathlength	10 mm
Reference product	SUPELCO
Purity	98.6%

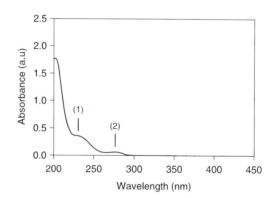

Peak n°	Wavelength (nm)	Absorbance (a.u.)
1	228.9	0.362
2	276.1	0.064

3.6.2. Di-butyl phthalate

General

Name	Di-butyl phthalate
CAS No.	84-74-2
Formula	$C_{16}H_{22}O_4$
Molecular weight	278.35 g
Solubility (H_2O)	11.2 mg/L (20°C)
Refractive index	1.49

Spectra acquisition

Solvent	MetOH/H_2O (0.2% v/v)
Concentration	10.2 mg/L
Pathlength	10 mm
Reference product	PROLABO
Purity	P.A.

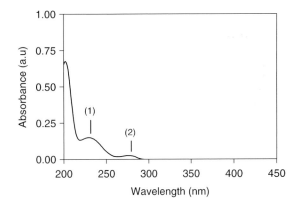

Peak n°	Wavelength (nm)	Absorbance (a.u.)
1	229.6	0.147
2	275.0	0.026

3.6.3. Di-ethyl phthalate

General

Name	Di-ethyl phthalate
CAS No.	84-66-2
Formula	$C_{12}H_{14}O_4$
Molecular weight	222.24 g
Solubility (H_2O)	1.080 g/L (25°C)
Refractive index	1.50

Spectra acquisition

Solvent	MetOH/H_2O (0.2% v/v)
Concentration	10.2 mg/L
Pathlength	10 mm
Reference product	SIGMA
Purity	99%

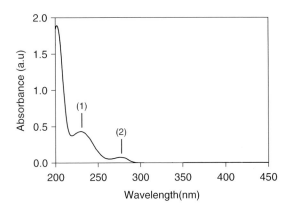

Peak n°	Wavelength (nm)	Absorbance (a.u.)
1	229.7	0.435
2	275.7	0.076

3.6.4. Potassium hydrogen phthalate

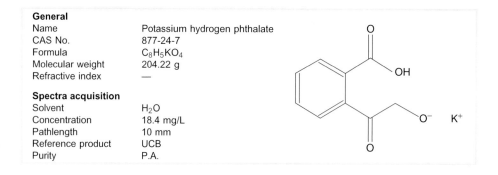

General
Name   Potassium hydrogen phthalate
CAS No.   877-24-7
Formula $C_8H_5KO_4$
Molecular weight 204.22 g
Refractive index —

Spectra acquisition
Solvent H_2O
Concentration   18.4 mg/L
Pathlength   10 mm
Reference product UCB
Purity   P.A.

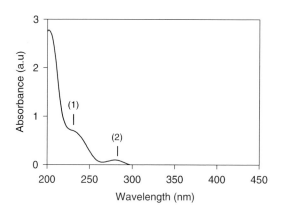

Peak n°	Wavelength (nm)	Absorbance (a.u.)
1	229.2	0.711
2	281.2	0.140

3.7. Surfactants

3.7.1. Alkyl diphenyloxide disulphonate, disodium salt

General
Name   Alkyl diphenyloxide disulphonate, disodium salt
CAS No.   28519-02-0/25167-32-2
Formula   —
Molecular weight   569 g
Refractive index   —

Spectra acquisition
Solvent H_2O
Concentration   50.0 mg/L
Pathlength   10 mm
Reference product   RHODIA (Rhodacal DSB®)
Purity   —

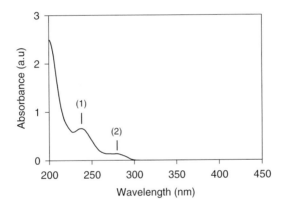

Peak n°	Wavelength (nm)	Absorbance (a.u.)
1	236.9	0.661
2	271.2	0.138

3.7.2. Dodecyl benzene sulphonate

General
Name   Dodecyl benzene sulphonate
CAS No.   25155-30-0
Formula $C_{18}H_{29}NaO_3S$
Molecular weight   325.49 g
Refractive index   —

Spectra acquisition
Solvent H_2O
Concentration   40.8 mg/L
Pathlength   10 mm
Reference product   FLUKA
Purity   Technic

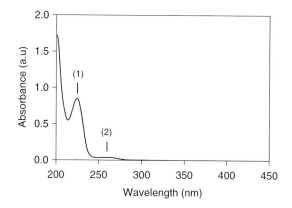

Peak n°	Wavelength (nm)	Absorbance (a.u.)
1	224.8	0.849
2	256.1	0.038

3.7.3. Nonyl phenol ethoxylate

General

Name	Nonyl phenol ethoxylate
CAS No.	68412-53-3
Formula	—
Molecular weight	—
Refractive index	—

Spectra acquisition

Solvent	H_2O
Concentration	50.0 mg/L
Pathlength	10 mm
Reference product	RHODIA (Rhodafac RE-610®)
Purity	—

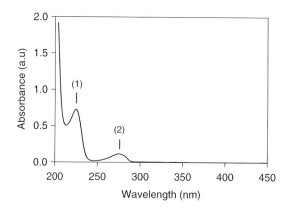

Peak n°	Wavelength (nm)	Absorbance (a.u.)
1	225.1	0.725
2	275.0	0.116

3.7.4. Octyl phenol ethoxylate

General

Name	Octyl phenol ethoxylate
CAS No.	9002-96-1
Formula	—
Molecular weight	—
Refractive index	—

Spectra acquisition

Solvent	H_2O
Concentration	50.0 mg/L
Pathlength	10 mm
Reference product	RHODIA (Igepal CA-630®)
Purity	—

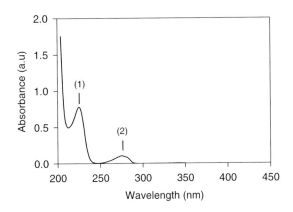

Peak n°	Wavelength (nm)	Absorbance (a.u.)
1	225.1	0.780
2	275.0	0.106

3.7.5. Sodium-N-methyl-N-oleoyl-taurate

General
Name	Sodium-N-methyl-N-oleoyl-taurate
CAS No.	137-20-2
Formula	—
Molecular weight	—
Refractive index	—

Spectra acquisition
Solvent	H_2O
Concentration	50.0 mg/L
Pathlength	10 mm
Reference product	RHODIA (Geropon T-77®)
Purity	—

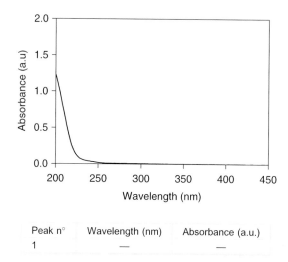

Peak n°	Wavelength (nm)	Absorbance (a.u.)
1	—	—

3.8. Pesticides

3.8.1. Atrazine

General
Name	Atrazine
CAS No.	1912-24-9
Formula	$C_8H_{14}ClN_5$
Molecular weight	215.68
Solubility (H_2O)	70 mg/L (20°C)
Refractive index	—

Spectra acquisition
Solvent	EtOH/H_2O (1% v/v)
Concentration	10.0 mg/L
Pathlength	10 mm
Reference product	RDH
Purity	Pestanal (98%)

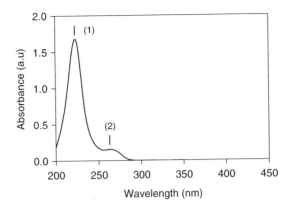

Peak n°	Wavelength (nm)	Absorbance (a.u.)
1	224.2	1.682
2	264.7	0.156

3.8.2. 2-4-Dichlorophenoxy acetic acid (2-4 D)

General

Name	2-4-Dichlorophenoxy acetic acid
CAS No.	94-75-7
Formula	$C_8H_6Cl_2O_3$
Molecular weight	221.04 g
Solubility (H_2O)	0.6 g/L at 20°C
Refractive index	—

Spectra acquisition

Solvent	MetOH/H_2O (0.3% v/v)
Concentration	21.1 mg/L
Pathlength	10 mm
Reference product	ALLTECH
Purity	99%
Remark	Spectra at pH = 4

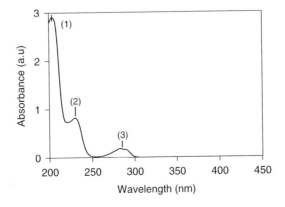

Peak n°	Wavelength (nm)	Absorbance (a.u.)
1	203.9	2.902
2	229.8	0.825
3	282.5	0.185

3.8.3. Dinoterb

General

Name	2-(1,1-Dimethyl-ethyl) 4,6-dinitrophenol
CAS No.	1420-07-1
Formula	$C_{10}H_{12}N_2O_5$
Molecular weight	240.21
Solubility (H_2O)	4.5 mg/L (20°C, pH $= 5$)
Refractive index	—

Spectra acquisition

Solvent	MetOH/H_2O (2.6% v/v)
Concentration	50.7 mg/L
Pathlength	10 mm
Reference product	RDH
Purity	Pestanal (99%)

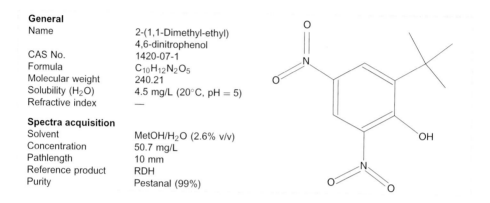

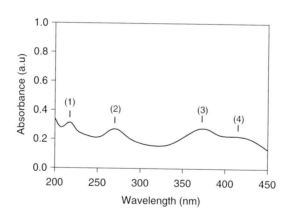

Peak n°	Wavelength (nm)	Absorbance (a.u.)
1	217.6	0.313
2	269.5	0.271
3	374.5	0.278
4	408.7	0.225

3.8.4. Diuron

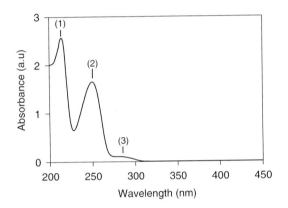

General

Name	Diuron
CAS No.	330-54-1
Formula	$C_9H_{10}Cl_2N_2O$
Molecular weight	233.09
Solubility (H_2O)	42 mg/L (25°C)
Refractive index	—

Spectra acquisition

Solvent	CH_3CN + MetOH (1 + 1) in H_2O (0.2% v/v)
Concentration	22.9 mg/L
Pathlength	10 mm
Reference product	SIGMA
Purity	98% min.

Peak n°	Wavelength (nm)	Absorbance (a.u.)
1	214.0	2.566
2	249.7	1.657
3	280.6	0.107

3.8.5. Hexazinone

General

Name	Hexazinone
CAS No.	51235-04-2
Formula	$C_{12}H_{20}N_4O_2$
Molecular weight	252.31
Solubility (H_2O)	33 g/L (20°C)
Refractive index	—

Spectra acquisition

Solvent	MetOH/H_2O (0.4% v/v)
Concentration	16.8 mg/L
Pathlength	10 mm
Reference product	SUPELCO
Purity	Etalon

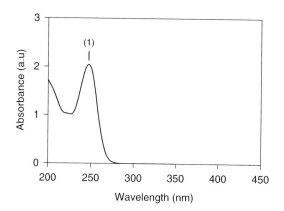

Peak n°	Wavelength (nm)	Absorbance (a.u.)
1	247.9	2.045

3.8.6. Paraquat

General

Name	Paraquat dichloride
CAS No.	1910-42-5
Formula	$C_{12}H_{14}Cl_2N_2$
Molecular weight	257.16
Solubility (H_2O)	>100 g/L (20°C)
Refractive index	—

Spectra acquisition

Solvent	H_2O
Concentration	31.1 mg/L
Pathlength	10 mm
Reference product	SIGMA
Purity	Etalon

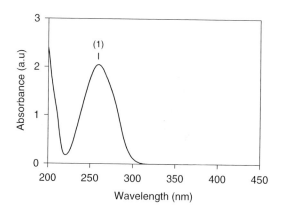

Peak n°	Wavelength (nm)	Absorbance (a.u.)
1	259.1	2.048

3.8.7. Parathion

General

Name	Parathion
CAS No.	56-38-2
Formula	$C_{10}H_{14}NO_5PS$
Molecular weight	291.26
Solubility (H_2O)	24 mg/L (24°C)
Refractive index	1.53

Spectra acquisition

Solvent	MetOH/H_2O (0.3% v/v)
Concentration	18.6 mg/L
Pathlength	10 mm
Reference product	RDH
Purity	Pestanal (98%)

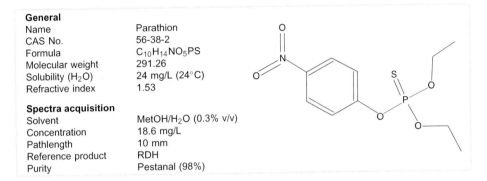

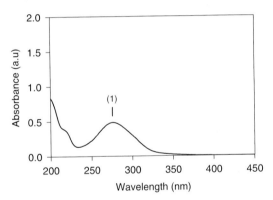

Peak n°	Wavelength (nm)	Absorbance (a.u.)
1	276.9	0.487

3.8.8. Simazine

General

Name	Simazine
CAS No.	122-34-9
Formula	$C_7H_{12}ClN_5$
Molecular weight	201.66
Solubility (H_2O)	5 mg/L (20°C)
Refractive index	—

Spectra acquisition

Solvent	EtOH/H_2O (1% v/v)
Concentration	4.9 mg/L
Pathlength	10 mm
Reference product	RDH
Purity	Pestanal (99%)

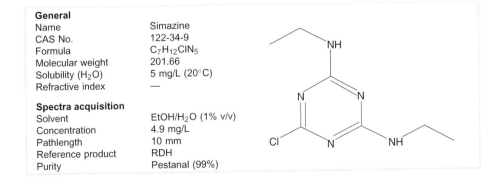

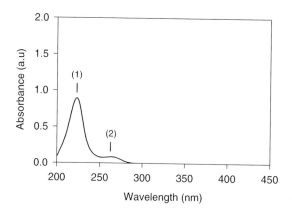

Peak n°	Wavelength (nm)	Absorbance (a.u.)
1	223.9	0.894
2	264.8	0.085

3.8.9. Terbutryn

General

Name	Terbutryn
CAS No.	886-50-0
Formula	$C_{10}H_{19}N_5S$
Molecular weight	241.36
Solubility (H_2O)	25 mg/L (20°C)
Refractive index	—

Spectra acquisition

Solvent	MetOH/H_2O (1% v/v)
Concentration	17.2 mg/L
Pathlength	10 mm
Reference product	RDH
Purity	Pestanal (98%)

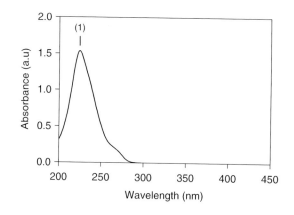

Peak n°	Wavelength (nm)	Absorbance (a.u.)
1	225.7	1.537

3.9. Polycyclic aromatic hydrocarbons

3.9.1. Acenaphthene

General

Name	Acenaphthene
CAS No.	83-32-9
Formula	$C_{12}H_{10}$
Molecular weight	154.20
Solubility (H_2O)	3.9 mg/L (20°C)
Refractive index	1.60

Spectra acquisition

Solvent	EtOH/H_2O (0.5% v/v)
Concentration	3.1 mg/L
Pathlength	10 mm
Reference product	ALDRICH
Purity	99%

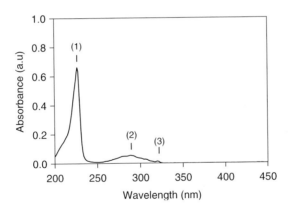

Peak n°	Wavelength (nm)	Absorbance (a.u.)
1	225.7	0.663
2	288.4	0.054
3	320.1	0.014

3.9.2. Acenaphthylene

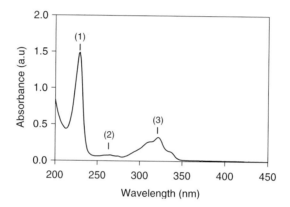

General

Name	Acenaphthylene
CAS No.	208-96-8
Formula	$C_{12}H_8$
Molecular weight	152.19
Solubility (H_2O)	16.3 mg/L (20°C)
Refractive index	—

Spectra acquisition

Solvent	EtOH/H_2O (0.2% v/v)
Concentration	5.2 mg/L
Pathlength	10 mm
Reference product	SUPELCO
Purity	98%

Peak n°	Wavelength (nm)	Absorbance (a.u.)
1	228.7	1.494
2	264.8	0.079
3	321.7	0.328

3.9.3. Anthracene

General

Name	Anthracene
CAS No.	120-12-7
Formula	$C_{14}H_{10}$
Molecular weight	178.23
Solubility (H_2O)	Not soluble
Refractive index	—

Spectra acquisition

Solvent	EtOH/CH_3CN (0.5% v/v)
Concentration	3.9 mg/L
Pathlength	10 mm
Reference product	ALDRICH
Purity	99%

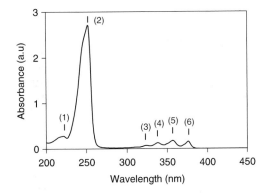

Peak n°	Wavelength (nm)	Absorbance (a.u.)
1	221.4	0.279
2	252.2	2.722
3	324.1	0.072
4	338.1	0.130
5	356.8	0.179
6	376.0	0.159

3.9.4. Benzo(a)anthracene

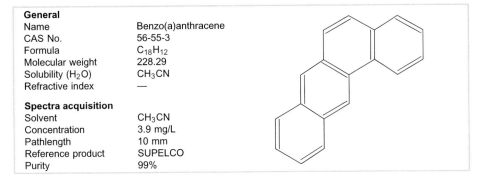

General
Name	Benzo(a)anthracene
CAS No.	56-55-3
Formula	$C_{18}H_{12}$
Molecular weight	228.29
Solubility (H_2O)	CH_3CN
Refractive index	—

Spectra acquisition
Solvent	CH_3CN
Concentration	3.9 mg/L
Pathlength	10 mm
Reference product	SUPELCO
Purity	99%

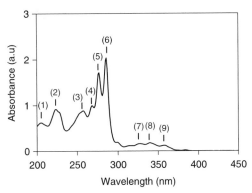

Peak n°	Wavelength (nm)	Absorbance (a.u.)
1	205.2	0.627
2	223.6	0.918
3	257.5	0.884
4	268.3	0.995
5	276.7	1.716
6	285.7	2.044
7	327.0	0.160
8	339.3	0.177
9	357.6	0.122

3.9.5. Benzo(a)pyrene

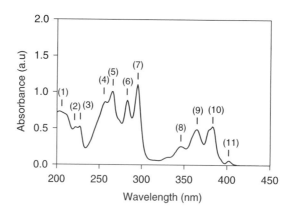

General

Name	Benzo(a)pyrene
CAS No.	50-32-8
Formula	$C_{20}H_{12}$
Molecular weight	252.31
Solubility (H_2O)	Not soluble
Refractive index	—

Spectra acquisition

Solvent	CH_3CN
Concentration	4.8 mg/L
Pathlength	10 mm
Reference product	SUPELCO
Purity	99%

Peak n°	Wavelength (nm)	Absorbance (a.u.)
1	203.1	0.732
2	222.2	0.518
3	226.8	0.526
4	256.8	0.865
5	265.4	1.011
6	283.0	0.887
7	295.0	1.105
8	346.2	0.256
9	364.7	0.492
10	383.1	0.534
11	402.0	0.061

3.9.6. Benzo(b)fluoranthene

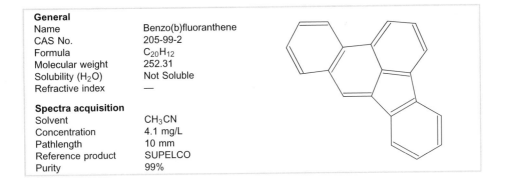

General
Name — Benzo(b)fluoranthene
CAS No. — 205-99-2
Formula — $C_{20}H_{12}$
Molecular weight — 252.31
Solubility (H_2O) — Not Soluble
Refractive index — —

Spectra acquisition
Solvent — CH_3CN
Concentration — 4.1 mg/L
Pathlength — 10 mm
Reference product — SUPELCO
Purity — 99%

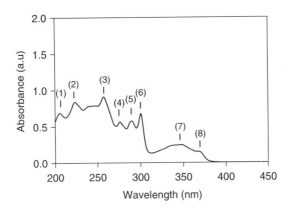

Peak n°	Wavelength (nm)	Absorbance (a.u.)
1	206.0	0.690
2	223.7	0.839
3	257.4	0.913
4	276.1	0.565
5	289.6	0.579
6	300.3	0.682
7	347.9	0.249
8	367.2	0.152

3.9.7. Benzo(g,h,i)perylene

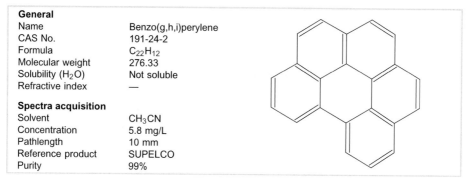

General

Name	Benzo(g,h,i)perylene
CAS No.	191-24-2
Formula	$C_{22}H_{12}$
Molecular weight	276.33
Solubility (H_2O)	Not soluble
Refractive index	—

Spectra acquisition

Solvent	CH_3CN
Concentration	5.8 mg/L
Pathlength	10 mm
Reference product	SUPELCO
Purity	99%

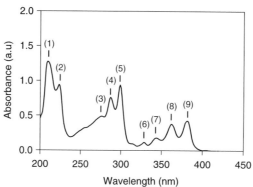

Peak n°	Wavelength (nm)	Absorbance (a.u.)
1	209.2	1.277
2	223.1	0.947
3	276.1	0.491
4	287.0	0.761
5	298.9	0.941
6	328.7	0.117
7	343.4	0.182
8	362.2	0.380
9	381.4	0.427

3.9.8. Benzo(k)fluoranthene

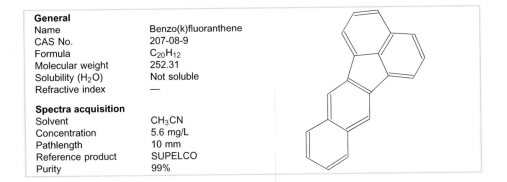

General	
Name	Benzo(k)fluoranthene
CAS No.	207-08-9
Formula	$C_{20}H_{12}$
Molecular weight	252.31
Solubility (H_2O)	Not soluble
Refractive index	—

Spectra acquisition	
Solvent	CH_3CN
Concentration	5.6 mg/L
Pathlength	10 mm
Reference product	SUPELCO
Purity	99%

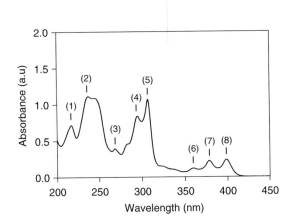

Peak n°	Wavelength (nm)	Absorbance (a.u.)
1	217.0	0.720
2	237.3	1.116
3	268.3	0.395
4	294.7	0.841
5	306.9	1.073
6	359.9	0.118
7	378.4	0.223
8	398.6	0.238

3.9.9.   Chrysene

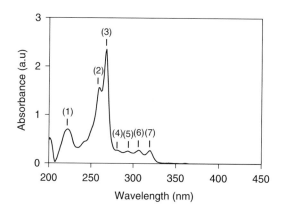

General
Name   Chrysene
CAS No.   218-01-9
Formula $C_{18}H_{12}$
Molecular weight   228.29
Solubility (H_2O)   Not soluble
Refractive index   —

Spectra acquisition
Solvent CH_3CN
Concentration   5.5 mg/L
Pathlength   10 mm
Reference product   SUPELCO
Purity   99%

Peak n°	Wavelength (nm)	Absorbance (a.u.)
1	222.5	0.706
2	259.5	1.565
3	267.7	2.361
4	280.5	0.276
5	293.6	0.258
6	306.6	0.277
7	319.7	0.272

3.9.10. Dibenz(a,h)anthracene

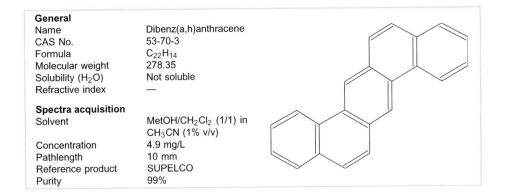

General

Name	Dibenz(a,h)anthracene
CAS No.	53-70-3
Formula	$C_{22}H_{14}$
Molecular weight	278.35
Solubility (H_2O)	Not soluble
Refractive index	—

Spectra acquisition

Solvent	MetOH/CH_2Cl_2 (1/1) in CH_3CN (1% v/v)
Concentration	4.9 mg/L
Pathlength	10 mm
Reference product	SUPELCO
Purity	99%

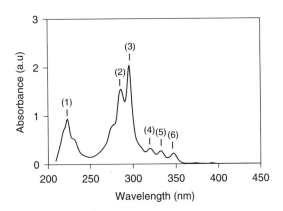

Peak n°	Wavelength (nm)	Absorbance (a.u.)
1	222.8	0.939
2	286.0	1.548
3	295.8	2.050
4	319.8	0.321
5	332.1	0.269
6	347.0	0.222

3.9.11. Fluoranthene

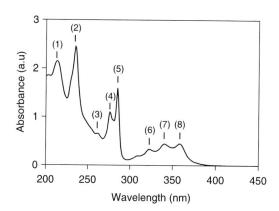

General	
Name	Fluoranthene
CAS No.	206-44-0
Formula	$C_{16}H_{10}$
Molecular weight	202.25
Solubility (H_2O)	Not soluble
Refractive index	—

Spectra acquisition	
Solvent	CH_3CN
Concentration	5.1 mg/L
Pathlength	10 mm
Reference product	SUPELCO
Purity	98%

Peak n°	Wavelength (nm)	Absorbance (a.u.)
1	202.8	1.860
2	213.4	2.151
3	235.2	2.459
4	276.1	1.102
5	285.1	1.589
6	322.6	0.337
7	340.5	0.446
8	358.1	0.451

3.9.12. Fluorene

General

Name	Fluorene
CAS No.	86-73-7
Formula	$C_{20}H_{12}$
Molecular weight	166.22
Solubility (H_2O)	Not soluble
Refractive index	1.31

Spectra acquisition

Solvent	EtOH/CH_3CN (1% v/v)
Concentration	6.7 mg/L
Pathlength	10 mm
Reference product	ALDRICH
Purity	98%

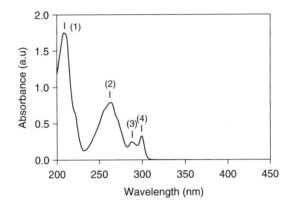

Peak n°	Wavelength (nm)	Absorbance (a.u.)
1	208.5	1.755
2	263.3	0.794
3	287.7	0.248
4	299.4	0.331

3.9.13. Indeno(1,2,3-cd)pyrene

General

Name	Indeno(1,2,3-cd)pyrene
CAS No.	193-39-5
Formula	$C_{22}H_{12}$
Molecular weight	267.33
Solubility (H_2O)	Not soluble
Refractive index	—

Spectra acquisition

Solvent	CH_3CN
Concentration	5.7 mg/L
Pathlength	10 mm
Reference product	SUPELCO
Purity	99%

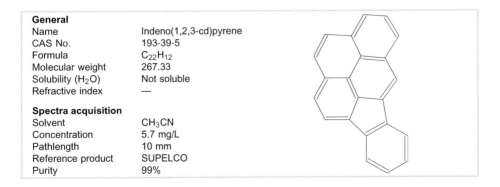

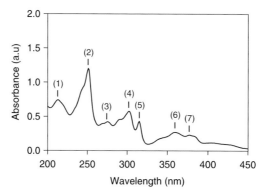

Peak n°	Wavelength (nm)	Absorbance (a.u.)
1	213.0	0.743
2	251.2	1.203
3	275.5	0.425
4	302.1	0.579
5	314.9	0.432
6	359.5	0.269
7	376.9	0.230

3.9.14. Naphthalene

General
Name	Naphthalene
CAS No.	91-20-3
Formula	$C_{10}H_8$
Molecular weight	128.17
Solubility (H_2O)	Not soluble
Refractive index	1.59

Spectra acquisition
Solvent	CH_3CN
Concentration	5.8 mg/L
Pathlength	10 mm
Reference product	ALDRICH
Purity	99%

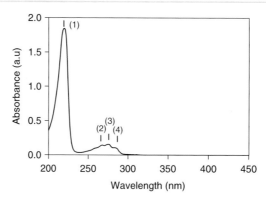

Peak n°	Wavelength (nm)	Absorbance (a.u.)
1	219.6	1.846
2	267.6	0.144
3	275.5	0.157
4	283.0	0.107

3.9.15. Phenanthrene

General
Name	Phenanthrene
CAS No.	85-01-8
Formula	$C_{14}H_{10}$
Molecular weight	178.29
Solubility (H_2O)	Not soluble
Refractive index	1.59

Spectra acquisition
Solvent	CH_3CN
Concentration	5.2 mg/L
Pathlength	10 mm
Reference product	MERCK
Purity	>97%

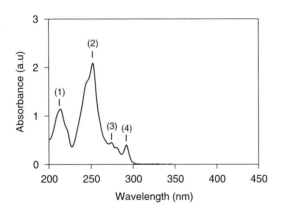

Peak n°	Wavelength (nm)	Absorbance (a.u.)
1	213.8	1.147
2	252.1	2.092
3	273.9	0.448
4	291.9	0.403

3.9.16. Pyrene

General	
Name	Pyrene
CAS No.	129-00-0
Formula	$C_{16}H_{10}$
Molecular weight	202.25
Solubility (H_2O)	Not soluble
Refractive index	—
Spectra acquisition	
Solvent	CH_3CN
Concentration	4.7 mg/L
Pathlength	10 mm
Reference product	MERCK
Purity	>97%

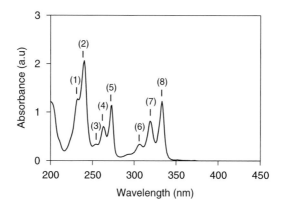

Peak n°	Wavelength (nm)	Absorbance (a.u.)
1	232.1	1.278
2	239.8	2.065
3	254.0	0.341
4	263.1	0.706
5	272.6	1.165
6	306.6	0.343
7	319.2	0.821
8	333.0	1.229

3.10. Solvents

3.10.1. Acetone

General
Name   Acetone
CAS No.   67-64-1
Formula C_3H_6O
Molecular weight   58.08 g
Solubility (H_2O)   Miscible
Refractive index   1.36

Spectra acquisition
Solvent H_2O
Concentration   1.565 g/L
pH   6.2
Pathlength   10 mm
Reference product CARLO ERBA
Purity   P. spectro (RS)

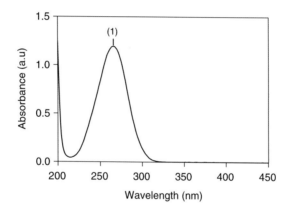

Peak n°	Wavelength (nm)	Absorbance (a.u.)
1	266.1	1.195

3.10.2. Acetonitrile

General
Name   Acetonitrile
CAS No.   75-05-8
Formula C_2H_3N
Molecular weight   41.05
Solubility (H_2O)   Miscible
Refractive index   1.36

Spectra acquisition
Concentration   Pure
Pathlength   10 mm
Reference product CARLO ERBA
Purity   RS-PLUS 99.9%

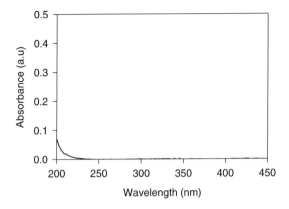

Peak n°	Wavelength (nm)	Absorbance (a.u.)
1	—	—

3.10.3. Ethanol

General

Name	Ethanol
CAS No.	64-17-5
Formula	C_2H_5O
Molecular weight	46.07
Solubility (H_2O)	Miscible
Refractive index	1.36

Spectra acquisition

Concentration	Pure
Pathlength	10 mm
Reference product	CARLO ERBA
Purity	RS 99.9%

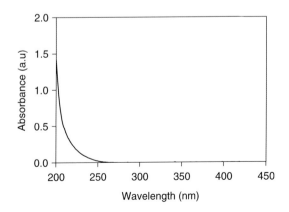

Peak n°	Wavelength (nm)	Absorbance (a.u.)
1	—	—

3.10.4. Hexane

General
Name	n-Hexane
CAS No.	110-54-3
Formula	C_6H_{14}
Molecular weight	86.18
Solubility (H_2O)	9.5 mg/L (20°C)
Refractive index	1.36

Spectra acquisition
Solvent	H_2O
Concentration	Pure
Pathlength	10 mm
Reference product	CARLO ERBA
Purity	RS 98%

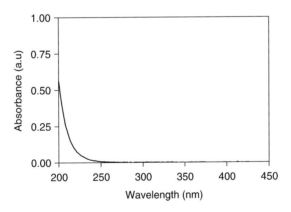

Peak n°	Wavelength (nm)	Absorbance (a.u.)
1	—	—

3.11. Inorganic compounds

3.11.1. Ammonium chloride

General	
Name	Ammonium chloride
CAS No.	12125-02-9
Formula	NH_4Cl
Molecular weight	53.49
Solubility (H_2O)	370 g/L (20°C)

Spectra acquisition	
Solvent	H_2O
Concentration	9554.0 mg/L
Pathlength	10 mm
Reference product	FLUKA
Purity	>99%

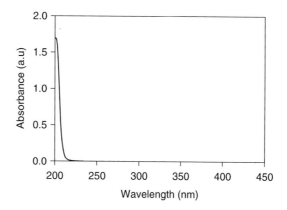

Peak n°	Wavelength (nm)	Absorbance (a.u.)
1	—	—

3.11.2. Hydrogen peroxide

General	
Name	Hydrogen peroxide
CAS No.	7722-84-1
Formula	H_2O_2
Molecular weight	34.01
Solubility (H_2O)	Miscible (20°C)

Spectra acquisition	
Solvent	H_2O
Concentration	342.0 mg/L
Pathlength	10 mm
Reference product	PROLABO (30% stabilised)
Purity	RP Normapur

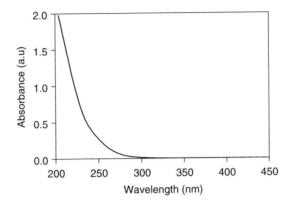

Peak n°	Wavelength (nm)	Absorbance (a.u.)
1	—	—

3.11.3. *Iodine*

General
Name	Iodine
CAS No.	7553-56-2
Formula	I_2
Molecular weight	253.8
Solubility (H_2O)	0.3 g/L (20°C)

Spectra acquisition
Solvent	H_2O
Concentration	50.3 mg/L
Pathlength	10 mm
Reference product	PROLABO
Purity	Titrated solution 0.1 N

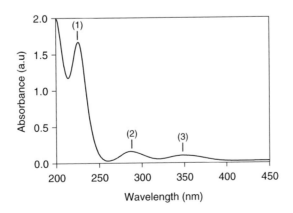

Peak n°	Wavelength (nm)	Absorbance (a.u.)
1	226.0	1.673
2	286.5	0.164
3	348.4	0.108

3.11.4. Potassium cyanide

General
Name	Potassium cyanide
CAS No.	151-50-8
Formula	KCN
Molecular weight	65.12
Solubility (H_2O)	680 g/L (20°C)

Spectra acquisition
Solvent	H_2O
Concentration	3648.0 mg/L
Pathlength	10 mm
Reference product	PROLABO
Purity	—

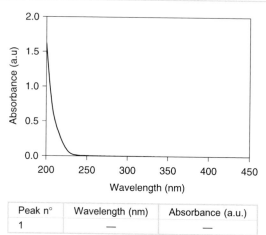

Peak n°	Wavelength (nm)	Absorbance (a.u.)
1	—	—

3.11.5. Potassium dichromate

General
Name	Potassium dichromate
CAS No.	7758-50-9
Formula	$K_2Cr_2O_7$
Molecular weight	294.18
Solubility (H_2O)	120 g/L (20°C)

Spectra acquisition
Solvent	H_2O
Concentration	53.07 mg/L
Pathlength	10 mm
Reference product	PROLABO
Purity	Normapur

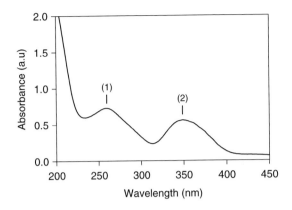

Peak n°	Wavelength (nm)	Absorbance (a.u.)
1	258.3	0.730
2	349.0	0.560

3.11.6. Potassium iodate

General
Name — Potassium iodate
CAS No. — 7758-05-6
Formula — KIO_3
Molecular weight — 214.0
Solubility (H_2O) — 47 g/L (20°C)

Spectra acquisition
Solvent — H_2O
Concentration — 55.6 mg/L
Pathlength — 10 mm
Reference product — CARLO ERBA
Purity — 99.9%

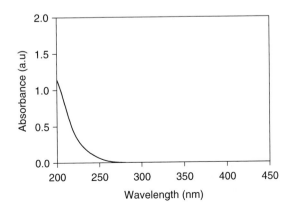

Peak n°	Wavelength (nm)	Absorbance (a.u.)
1	—	—

3.11.7. Potassium iodide

General
Name   Potassium iodide
CAS No.   7681-11-0
Formula   KI
Molecular weight   166.01
Solubility (H_2O) 1270 g/L (20°C)

Spectra acquisition
Solvent H_2O
Concentration   36.0 mg/L
Pathlength   10 mm
Reference product   MERCK
Purity   Suprapur

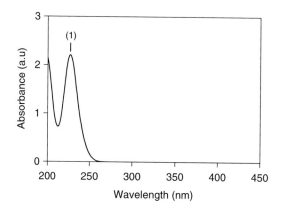

Peak n°	Wavelength (nm)	Absorbance (a.u.)
1	227.0	2.210

3.11.8. Potassium metaperiodate

General
Name   Potassium metaperiodate
CAS No.   7790-21-8
Formula KIO_4
Molecular weight   230.00
Solubility (H_2O) 4.2 g/L (20°C)

Spectra acquisition
Solvent H_2O
Concentration   50.7 mg/L
Pathlength   10 mm
Reference product   LABOSI
Purity   Analypur

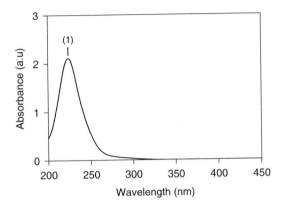

Peak n°	Wavelength (nm)	Absorbance (a.u.)
1	223.7	2.109

3.11.9. Potassium permanganate

General

Name	Potassium permanganate
CAS No.	7722-64-7
Formula	$KMnO_4$
Molecular weight	158.03
Solubility (H_2O)	70 g/L (20°C)

Spectra acquisition

Solvent	H_2O
Concentration	100.7 mg/L
Pathlength	10 mm
Reference product	UCB
Purity	Pure

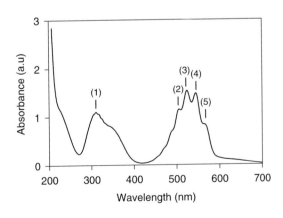

Peak n°	Wavelength (nm)	Absorbance (a.u.)
1	310.8	1.102
2	506.1	1.149
3	524.4	1.531
4	544.8	1.476
5	564.5	0.837

3.11.10. Sodium chlorate

General
Name	Sodium chlorate
CAS No.	7775-09-9
Formula	$NaClO_3$
Molecular weight	106.44
Solubility (H_2O)	1000 g/L (10°C)

Spectra acquisition
Solvent	H_2O
Concentration	2034.3 mg/L
Pathlength	10 mm
Reference product	RDH
Purity	99.5%

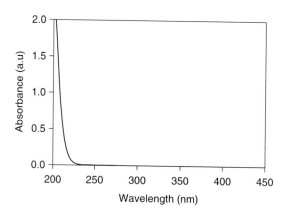

Peak n°	Wavelength (nm)	Absorbance (a.u.)
1	—	—

3.11.11. Sodium chromate

General

Name	Sodium chromate tetrahydrate
CAS No.	10034-82-9
Formula	$Na_2CrO_4, 4H_2O$
Molecular weight	161.97
Solubility (H_2O)	873 g/L (10°C)

Spectra acquisition

Solvent	H_2O
Concentration	56.4 mg/L
Pathlength	10 mm
Reference product	ALDRICH
Purity	99%

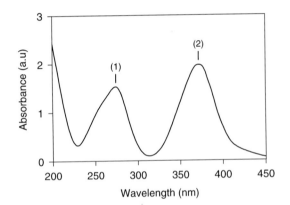

Peak n°	Wavelength (nm)	Absorbance (a.u.)
1	274.1	1.530
2	372.3	1.975

3.11.12. Sodium cyanide

General

Name	Sodium cyanide
CAS No.	143-33-9
Formula	NaCN
Molecular weight	49.01
Solubility (H_2O)	480 g/L (10°C)

Spectra acquisition

Solvent	H_2O
Concentration	2061.3 mg/L
Pathlength	10 mm
Reference product	PROLABO
Purity	—

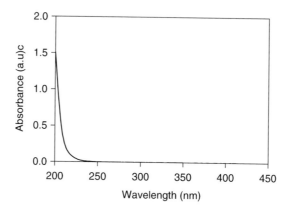

Peak n°	Wavelength (nm)	Absorbance (a.u.)
1	—	—

3.11.13.   Sodium hypochlorite

General	
Name	Sodium hypochlorite (in solution: "Eau de Javel")
CAS No.	7681-52-9
Formula	NaOCl
Molecular weight	74.44
Solubility (H_2O)	Very soluble
Spectra acquisition	
Solvent	H_2O
Concentration	794.2 mg/L (active chlorine)
Pathlength	10 mm
Reference product	JAVEL IDEAL
Purity	Commercial product

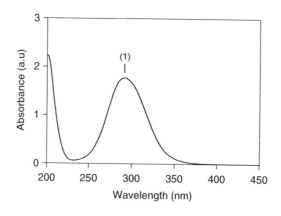

Peak n°	Wavelength (nm)	Absorbance (a.u.)
1	291.4	1.780

3.11.14. Sodium nitrate (low concentration)

General
Name	Sodium nitrate
CAS No.	7631-99-4
Formula	$NaNO_3$
Molecular weight	84.99
Solubility (H_2O)	880 g/L (20°C)

Spectra acquisition
Solvent	H_2O
Concentration	20.7 mg/L
Pathlength	10 mm
Reference product	FLUKA
Purity	>99%

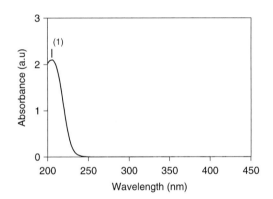

Peak n°	Wavelength (nm)	Absorbance (a.u.)
1	205.6	2.102

3.11.15. Sodium nitrate (high concentration)

General
Name	Sodium nitrate
CAS No.	7631-99-4
Formula	$NaNO_3$
Molecular weight	84.99
Solubility (H_2O)	880 g/L (20°C)

Spectra acquisition
Solvent	H_2O
Concentration	8864.0 mg/L
Pathlength	10 mm
Reference product	FLUKA
Purity	>99%

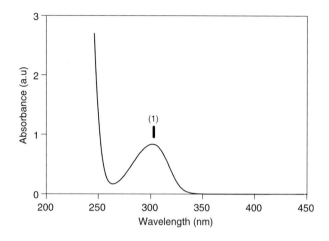

Peak n°	Wavelength (nm)	Absorbance (a.u.)
1	301.6	0.837

3.11.16. Sodium nitrite (low concentration)

General
Name   Sodium nitrite
CAS No.   7632-00-0
Formula $NaNO_2$
Molecular weight   69.00
Solubility (H_2O) 820 g/L (20°C)

Spectra acquisition
Solvent H_2O
Concentration   10.6 mg/L
Pathlength   10 mm
Reference product   CARLO ERBA
Purity   99%

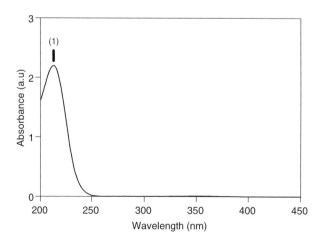

Peak n°	Wavelength (nm)	Absorbance (a.u.)
1	212.8	2.199

3.11.17. Sodium nitrite (high concentration)

General
Name   Sodium nitrite
CAS No.   7632-00-0
Formula $NaNO_2$
Molecular weight 69.00
Solubility (H_2O) 820 g/L (20°C)

Spectra acquisition
Solvent H_2O
Concentration   3007.0 mg/L
Pathlength   10 mm
Reference product CARLO ERBA
Purity   99%

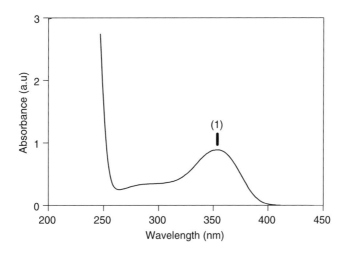

Peak n°	Wavelength (nm)	Absorbance (a.u.)
1	353.9	0.890

3.11.18.   Sodium tetraborate decahydrate

General
Name	Sodium tertraborate decahydrate (borax)
CAS No.	1303-96-4
Formula	$Na_2B_4O_7, 10H_2O$
Molecular weight	381.37
Solubility (H_2O)	50 g/L (20°C)

Spectra acquisition
Solvent	H_2O
Concentration	10368.1 mg/L
Pathlength	10 mm
Reference product	ALDRICH
Purity	ACS

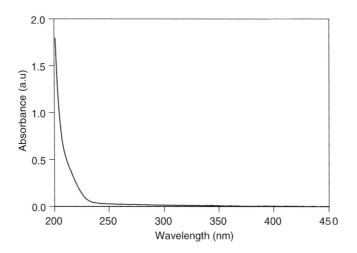

Peak n°	Wavelength (nm)	Absorbance (a.u.)
1	—	—

References

1. H.H. Perkampus, *UV–Vis Atlas of Organic Compounds,* 2nd Edition, Weinheim, VCH (1992).
2. A. Noelle, G.K. Hartmann, D. Lary, W.-U. Palm, A.-C. Vandaele, R.P. Wayne, R. Wu, *UV-Visible Spectra Data Base*, 4th Edition on CD-ROM (2005) (see www.uv-spectra.de).

INDEX